Lecture Notes in Physics

For information about Vols. 1–23, please contact your bookseller or Springer-Verlag.

Lecture Notes in Physics

Edited by J. Ehlers, München, K. Hepp, Zürich
R. Kippenhahn, München, H. A. Weidenmüller, Heidelberg
and J. Zittartz, Köln
Managing Editor: W. Beiglböck, Heidelberg

101

A. Martin-Löf

Statistical Mechanics and the Foundations of Thermodynamics

Springer-Verlag Berlin Heidelberg GmbH 1979

Author

Anders Martin-Löf
Institutet för
Försäkringsmatematik
och Matematisk Statistik
Stockholms Universitet
Hagagatan 23, Box 6701
113 85 Stockholm
Schweden

ISBN 978-3-540-09255-1   ISBN 978-3-540-35293-8 (eBook)

DOI 10.1007/978-3-540-35293-8

Library of Congress Cataloging in Publication Data
Martin-Löf, Anders, 1940- Statistical mechanics and the foundations of thermodynamics.
(Lecture notes in physics ; 101) Bibliography: p. Includes index. 1. Statistical thermo-
dynamics. I. Title. II. Series.
QC311.5.M28   536'.7   79-15289

© by Springer-Verlag Berlin Heidelberg 1979

Originally published by Springer-Verlag Berlin Heidelberg New York in 1979

Preface

These lectures present an introduction to classical statistical mechanics
and its relation to thermodynamics. They are intended to bridge the gap
between the treatment of the subject in the physics text books and in the
modern presentations of mathematically rigorous results. (So it tries to
supply many of the facts that are to be found between the lines both in
Landau-Lifshitz´ Statistical Physics and Ruelles´ Statistical Mechanics).

We have put some emphasis on getting a detailed and logical presentation
of the foundations of thermodynamics based on the maximum entropy principles
which govern the values taken by macroscopic variables according to the
laws of large numbers. These can be given a satisfactory formulation using
the limits of the basic thermodynamic functions established in the modern
work on rigorous results.

The treatment is reasonably self contained both concerning the physics and
mathematics needed. No knowledge of quantum mechanics is presupposed.
Since we present mathematical proofs of many technical facts about the
thermodynamic functions perhaps the treatment is most digestive for the
mathematically inclined reader who wants to understand the physics of the
subject, but it is hoped that the treatment of the basis of thermodynamics
is also clarifying to physically inclined readers.

Contents

Introduction

Statistical Mechanics is the branch of theoretical physics which investigates
and tries to relate macroscopic properties of systems of many interacting
microscopic subsystems (atoms, molecules etc.) to what is known about the
laws of the interactions of these atomic parts. The macroscopic description
uses a few ($<< 10^{23}$) variables which in general do not indicate the states
of individual atoms, and one looks for relations between these variables
and tries to describe their evolution with a few (differential) equations
only containing these variables (macroscopic causality).

These relations are found by considering the state variables of the individual subsystems as random variables whose distributions contain a few parameters. The average values of the macroscopic variables then only depend on
these few parameters, and the fact that these mean values well indicate
the actual values of the variables depend on the fact that "laws of large
numbers" hold for them with overwhelming precision. An example of such a
macroscopic theory is thermodynamics which e.g. deals with the equations of
state for gases, liquids, chamical systems in equilibrium, where typical
macroscopic state variables are: pressure, densities, temperature, heat
content etc. An important problem in this area is to explain phase transitions between gas-liquid liquid-solid, etc. and to understand if such
phenomena can be explained only from rather general assumptions about the
atomic interactions.

Areas where the time evolution of macroscopic systems is studied are
hydrodynamics, diffusion, heat conduction, kinetic theory of gases. Here
one needs many more state variables (density- and velocity distributions in
space e.g.) but one still has a drastic reduction from the microscopic
description, and one considers the evolution on a time scale which is very
long compared to that of the atomic motions. A fundamental problem is to
understand how the equations of motion for the macroscopic variables, which
can often be written down intuitively, can be derived from the molecular

dynamics and why macroscopic causality holds. These dynamic problems are
much more difficult than the equilibrium problems. In the following we
will consider mostly the latter type of problems.

The theory not only gives relations between the average values of the macroscopic variables but also describes their fluctuations (which are small).
For these in general central limit theorems are valid, because one has sums
of weakly dependent variables, and one obtains Gaussian distributions.
Examples: the theory of Brownian motion and density fluctuations in a gas
(which gives rise to the blue color of the sky).

In these lectures we shall first introduce the probability distributions,
"ensembles", appropriate for describing systems in equilibrium and consider
some of their basic physical applications. We also discuss the problem of
approach to equilibrium and the problem of how irreversibility comes into the
dynamics in the context of the so called Ehrenfest urn model , which is one
of the simplest systems where these questions can be studied quite explicitly.

We then give a detailed description of how the law of large numbers for macrovariables in equilibrium is derived from the fact that entropy is an extensive quantity in the "thermodynamic limit" $N,E,V \to \infty$ with $\frac{N}{V} = n =$ particle
density, $\frac{E}{V} = e =$ energy density. The discussion is based very much on the
fundamental convexity properties of the entropy and other thermodynamic
functions and shows how the values of the macrovariables are determined by
appropriate variational principles, which give rise to the rules for thermodynamical equilibrium. In this context we show that one can naturally see
how to split the energy changes in a thermodynamical process into work and
heat and show how the first and second laws of thermodynamics are derived
from the rules for thermodynamical equilibrium. We have elaborated the derivation of thermodynamics in detail because we feel that it is often insufficient-
ly treated in the physical text books and taken for granted in the more modern
treatments of "rigorous results". We feel it is satisfactory that the estab-
lishment of the limit of the thermodynamic functions achieved in the modern
development of the mathematical aspects of statistical mechanics allows a
more general and logically clear presentation of the bases of thermodynamics.
The proofs presented of the thermodynamic limit are somewhat elaborated ver-
sions of those in the litterature. Finally we present the basic facts about
fluctuation theory.
Since the theory deals with atomic phenomena it is often important to use
quantum mechanics to describe these, we shall however mostly deal with
systems described by classical mechanics, so we will not assume that the
reader is familiar with quantum mechanics. (It was in this area that
quantum mechanical effects were first encountered: Planck's radiation law,
the specific heat of solids and gases e.g.)

1. Statistical description of systems in classical mechanics

The prototype of such a systems is a gas (liquid) of N identical atoms
described as mass points moving according to Newton's equations under
the influence of pairwise interactions between the particles and collissions
between the particles and the walls of the container:

The positions are $(q_1, q_2, \ldots q_N)$ $q_i \in \Lambda \subset R^3$ Λ = the container. Newtons
equations are:

$$m\ddot{q}_i = - \sum_{j \neq i} \text{grad } V(q_i - q_j) \qquad \text{for } q_i \in \text{int } \Lambda$$

$$+ \text{ the law of reflextion} \qquad \text{for } q_i \in \partial \Lambda .$$

$V(q)$ is the potential of the interaction between a pair whose positions
differ by $q \in R^3$, $V(-q) = V(q)$. Typically V is repulsive at short
distances and attractive at large ones:

These equations can be written in Hamilton's canonical form like for any
mechanical system, and it will be very useful as we will see:

Introduce the momenta of the particles $p_i = m\dot{q}_i$ and the total energy as
a function of q_i, p_i :

$$H(p,q) = \sum_i \frac{m|\dot{q}_i|^2}{2} + \sum_{i<j} V(q_i - q_j) = \sum_i \frac{|p_i|^2}{2m} + \sum_{i<j} V(q_i - q_j)$$

Then we have:

$$\frac{\partial H}{\partial p_i} = \frac{p_i}{m} = \dot{q}_i$$

$$\frac{\partial H}{\partial q_i} = \sum_{i<j} \text{grad } V(q_i - q_j) - \sum_{j<i} \text{grad } V(q_j - q_i)$$

$$= \sum_{j \neq i} \text{grad } V(q_i - q_j) = - \dot{p}_i ,$$

which is the canonical form

$$\dot{q}_i = \frac{\partial H(p,q)}{\partial p_i}$$

$$\dot{p}_i = -\frac{\partial H(p,q)}{\partial q_i}$$

The mechanical theory of such a system consists in solving the system of differential equations with given initial values of (p,q). This is of course in general impossible for $N \approx 10^{23}$, and even if it were possible one could not hope to measure all initial values needed. Moreover one is not interested in such a detailed description in order to determine the average values of a few quantities over times long compared to the atomic ones. Therefore it is natural to try to treat all variables of whose initial values we have no precise information as random, i.e. to introduce a probability measure in the "phase space"

$$\Gamma = \{x = (p_1, q_1, \ldots, p_N, q_N)\} \subset R^{6N}$$

(Boltzmann, Gibbs, Maxwell)

This is often expressed by saying that we have an "ensemble" of infinitely many identical independent copies of the system whose initial values are distributed according to the given measure having a density $\rho(x)dx$. The hope is then that one can find a $\rho(x)$ such that averages with respect to this measure of interesting functions will give their observed values at least if their variances are small. If $\rho(x)$ is to describe a system in macroscopic equilibrium it ought to be chosen so that x_t becomes a stationary stochastic process. The equations of motion define a family of transformations $x_0 \to x_t = T_t(x_0)$ of $\Gamma \to \Gamma$, and stationarity means that ρ defines a stationary measure: $E(f(x_t)) = E(f(x_0))$ for all t and all integrable $f(x)$, which means that

$$\int f(x)\rho(x)dx = \int f(T_t(x))\rho(x)dx = \int f(y)\rho(T_{-t}(y)) \left| \frac{dT_{-t}(y)}{dy} \right| dy.$$

$\left| \dfrac{dT_t(y)}{dy} \right|$ is the Jacobian of the transformation $y \to T_t(y)$.

The remarkable fact is now that for canonical systems it is easy to find stationary measures:

Theorem 1 (Liouville's theorem)

The Lebesque measure $dx = dp_1 dq_1 \ldots dq_N$ in Γ is stationary, and more

generally, if $\rho(x)$ is a constant of the motion, i.e. if $\rho(x_t)$ is independent of t, then $\rho(x)dx$ defines a stationary measure.

Proof: From the above formula we see that it is enough to show that

$$J_t(x) = \left| \frac{dT_t(x)}{dx} \right| = 1 \quad \text{for all} \quad t. \quad \text{Clearly} \quad J_o(x) = |I| = 1, \quad \text{so we show}$$

that $\dfrac{dJ_t(x)}{dt} = 0$ for all t.

$$J_{t+h}(x) = \left| \frac{dT_{t+h}(x)}{dx} \right| = \left| \frac{dT_h(T_t(x))}{dx} \right| = J_h(x_t) \cdot J_t(x), \quad \text{because the Jacobian}$$

of a composed transformation is the product of the Jacobian of the factors.
For h small we have $x_h = J_h(x) \approx x + h\dot{x}$ from the equations of motion

where $\dot{x} = \begin{bmatrix} -\dfrac{\partial H(p,q)}{\partial q} \\ \dfrac{\partial H(p,q)}{\partial p} \end{bmatrix} \quad \text{if} \quad x = \begin{bmatrix} p \\ q \end{bmatrix}.$

Hence $J_h(x) \approx |I + hK|$, where K is given by

$$K = \begin{bmatrix} -\dfrac{\partial^2 H}{\partial q \partial p} & -\dfrac{\partial^2 H}{\partial q \partial q} \\ \dfrac{\partial^2 H}{\partial p \partial p} & \dfrac{\partial^2 H}{\partial p \partial q} \end{bmatrix}$$

We then have

$$J_h(x) = 1 + hspK + o(h^2) = 1 + h \sum_i \left(-\frac{\partial^2 H}{\partial q_i \partial p_i} + \frac{\partial^2 H}{\partial p_i \partial q_i} \right) + o(h^2) =$$

$$= 1 + o(h^2),$$

so we see that

$$\frac{J_{t+h} - J_t}{h} \to o \quad \text{as} \quad h \to o, \quad \text{and} \quad \frac{dJ_t(x)}{dt} = 0.$$

In particular, since $H(x)$ is a constant of the motion, we see that ρ
defines a stationary measure if it depends only on $H(x)$:

$$\rho(x) = \rho(H(x)) .$$

For an isolated system like the one we have considered H(x) is a constant
of the motion, so if its value is known, H(x) = E, we know that the system
moves on the "energy shell" Γ_E defined by this equation. The "simplest"
choice of ρ in this case is to choose the measure μ induced on Γ_E by
Lebesque measure in Γ:

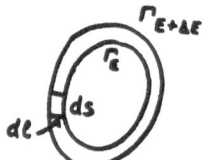

Put $d\mu = \text{const.} \dfrac{ds\,d\ell}{dE}$

ds = element of area on Γ_E

(The normal to Γ_E at x is grad H(x), and the thickness of an infinite-
simal shell is given by dE = |grad H|·dℓ). We have:

$$d\mu(x) = \frac{ds(x)}{\Omega'(E,N)\,|grad\ H(x)|\,N!} , \quad \text{with}$$

$$\Omega'_\Lambda (E,N) = \int_{\Gamma_E} \frac{ds}{|grad\ H|\,N!} = \frac{d}{dE} \int_{\substack{H(p,q)\leq E \\ q\in\Lambda^N}} \frac{dp\,dq}{N!} \equiv \frac{d\Omega_\Lambda(E,N)}{dE}$$

This is called the "microcanonical" measure (Gibbs). (The factor N! will
turn out to be natural later.)

If there are other known constants of the motion it could be natural to let
the microcanonical measure be defined on the lower dimensional manifold
defined by fixing their values, but in general H(x) is the only constant
known (besides the total momentum and angular momentum, but they will
automatically have the value zero in a large system, and any other fixed
value of these quantites can be reduced to zero by a suitable choice of
the coordinate system).

The choice of μ as a stationary measure was suggested by Boltzmann, who
saw it as a device for calculating time averages of interesting observables.
Such averages he considered to be the natural quantities to be compared to
observed values. In order to know that $\langle f \rangle = \lim_{T\to\infty} \frac{1}{T} \int_0^T f(x_t)\,dt$ a.s.

e.g. for all integrable f one needs to know that the system is ergodic
(Birkhoff's theorem). It is however only for a few very simplified systems

that it has been possible to establish ergodicity (Billiards, Sinai 1966)
so this motivation for the choice of μ is not yet available in general.
It is not necessary that the equality is valid for all such f. The physically interesting functions are often of a special kind, namely sums of
small contributions from the individual particles (or pairs etc.) which
are very many. For such f it is often true that $\langle (f - \langle f \rangle)^2 \rangle$ is
very small and then as pointed out by Khinchine in ref. 7 it is also true that

$$\frac{1}{T} \int_0^T f(x_t) dt \approx \langle f \rangle \qquad \text{when N is large. Also, if}$$

$\langle f(x_t) f(x_o) \rangle - \langle f(x_o) \rangle^2 \to 0$ sufficiently rapidly it is also true that
the time average converges to $\langle f \rangle$, so one should probably make more use
of the structure of the interesting observables in establishing the equality of the two averages. Even if it would be possible to show that a system
is not ergodic, i.e. that Γ_E can be decomposed into two invariant parts
with positive measure, it is quite possible that all interesting observables
are insensitive to this decomposition, and therefore ergodicity is not the
single criterion for whether the choice of measure is the appropriate one.

The fundamental problem of explaining why macroscopic systems not in equilibrium often quickly move towards equilibrium (described by equ. statistical mechanics) is not solved by showing that the system is ergodic with
μ as stationary measure. Here it is a question about the dynamics of
macroscopic variables on a time scale long compared to that of the atoms
starting from an initial state, which has extremely small μ-probability.
This problem occupied Boltzmann very much, and he invented his famous
equation to describe the evolution of the state of a gas of the type we
have described at least for low densities. In Boltzmanns description one
does not give an equation for the whole microscopic state $X_t \in \Gamma$, but
the state variable is a much more "coarse grained" quantity, namely the
density $f(p,q) dp dq$, $(p,q) \in R^6 = \gamma$. $f(p,q)$ is the particle density
in the state space of a single particle (γ-space), and it is considered to
be given on a microscopically long scale, $N \cdot f(p,q) dp dq$ is the number of
points in Γ whose coordinates $(p,q) \in dp dq$. Here $dp dq$ are "macroscopically infinitesimal", but "microscopically infinite", so
$N \cdot f \cdot dp \cdot dq \gg 1$, and the fluctuations in f are supposed to be small.
B´s equation is a deterministic differential equation for the evolution
of f, and for it he could show his H-theorem, which says that if one
starts from a $f_o(p,q)$ different from the equilibrium density f_∞ then
$f_t \to f_\infty$ as $t \to \infty$. f_∞ can be computed from μ described above.

This theorem and the question if equilibrium can be established gave rise
to controversies and contradictions. Zermelo´s "Wiederkehrseinwand" was
built on the observation that the microscopic equations of motion are
reversible, if time runs backwards one again gets a possible evolution.
B´s equation however is not, if one starts from f_∞ one does not come
back to f_0 after a long time, but f_∞ stays constant, so it appears
paradoxical that the system has suddenly lost its reversibility. Poincaré
also showed his recurrence theorem, which says that for every subset $A \subset \Gamma$
with $\mu(A) > o$ it is true for almost all $x \in A$ that if $x_0 = x$ then
$x_t \in A$ for some $t > o$. But this is not true for f_t, $f_t \to f_\infty$ and does
not come back near its initial value. Since the B-equation was not derive completely from the dynamics in Γ but written down with the aid
of Boltzmann´s intuition it was not clear how the paradoxes could be
explained.

The connection between the microscopic and macroscopic dynamics and the
problem how irreversibility can sneak into the equations can be illustrated
in a much simpler system where one can see what happens. This system is
the so called urnmodel which was introduced by Ehrenfest for this purpose:
Consider a "gas" of N particles labeled $1,2,\ldots,N$, distributed between
two urns U_0,U_1. The microscopic state $x \in T$ is given by a vector
$x = (x_1,\ldots,x_N)$ with $x_i = x$ iff particle no. $i \in U_x$, $x = 0,1$. The time
evolution takes place because once every time unit a particle, no. n, is
chosen at random and moved to the other urn, $x_n \to 1 - x_n$, etc. The system
hence forms a Markov chain with 2^N states, and it is easy to show that
the uniform distribution $\mu(x) = 2^{-N}$ is stationary, and that the process
is reversible under μ: $\mu(x)P(x,y) = \mu(y)P(y,x)$. It is also recurrent,
from every state one sooner or later comes to every other state. This
system corresponds to the mechanical system with the microscopic state
description and the microcanonical stationary measure. A macroscopic state
description can now e.g. consist in giving only the value of
$X_t = \sum_i x_i(t) = $ no. of particles in U_1, a drastic reduction in precision.
The induced distribution for X is easy to see.

$$p_n = P(X = n) = 2^{-N} \binom{N}{n} .$$

For such a macroscopic quantity the law of large numbers is hence valid
with great precision:

$$P(\left| \frac{X}{N} - \frac{1}{2} \right| > \epsilon) \le \text{const } e^{-N\left[h(\frac{1}{2}) - h(\frac{1}{2} + \epsilon)\right]} \to 0 \quad \text{as} \quad N \to \infty .$$

(Here $h(p)$ is the usual entropy function $h(p) = - p \log p - (1-p) \log(1-p)$.)
The fluctuations are also normal: $\sqrt{N} \left(\frac{X}{N} - \frac{1}{2}\right)$ has an approximatively normal
distribution $N(0, \frac{1}{2})$ as $N \to \infty$.

The macroscopic dynamics can also be studied exactly. X_t is also a Markov
chain with transition probabilities

$$\begin{cases} P_{n,n+1} = 1 - \frac{n}{N} \\ \\ P_{n,n-1} = \frac{n}{N} \end{cases} \qquad n = 0,1,\ldots,N$$

The forwards equation for $P_n(t) = P(X_t = n)$ is hence

$$P_n(t+1) = P_{n-1}(t) \left(1 - \frac{n-1}{N}\right) + P_{n+1}(t) \left(\frac{n+1}{N}\right) ,$$

and it is easy to see that the p_n above are the stationary probabilities.
Under these probabilities X_t is also reversible and recurrent, because
any state can be reached from any other state, and $P_n P_{n,m} = P_m P_{m,n}$,
as is easily checked. If we now consider X_t on a macroscopic scale, and
on a long time scale $t = N \cdot \tau$: $\frac{X_{N \cdot \tau}}{N} = f(\tau)$ we shall see that $f(\tau)$ will
vary completely deterministically when $N \to \infty$. To see this consider the
time evolution of $\langle f(\tau) \rangle = f_1(\tau)$ and $\langle f^2(\tau) \rangle = f_2(\tau)$, For one time
step $\Delta \tau = \frac{1}{N}$ we have:

$$\begin{cases} \left\langle f(\tau + \frac{1}{N}) \,\middle|\, f(\tau) = f \right\rangle = (1 - f)(f + \frac{1}{N}) + f(f - \frac{1}{N}) = f + \frac{(1 - 2f)}{N} \\ \\ \left\langle f^2(\tau + \frac{1}{N}) \,\middle|\, f(\tau) = f \right\rangle = (1 - f)(f + \frac{1}{N})^2 + f(f - \frac{1}{N})^2 = \\ \\ \qquad = f^2 + \frac{2f - 4f^2}{N} + 0(\frac{1}{N^2}) \end{cases}$$

So, taking averages of $f(\tau)$ we see that $f_1(\tau)$, $f_2(\tau)$ satisfy the
equations:

$$\begin{cases} f_1(\tau + \frac{1}{N}) = f_1(\tau) + \frac{1}{N}(1 - 2f_1(\tau)) \\ \\ f_2(\tau + \frac{1}{N}) = f_2(\tau) + \frac{2}{N}(f_1(\tau) - 2f_2(\tau)) + 0(\frac{1}{N^2}) , \end{cases}$$

and it is easy to show that as $N \to \infty$ f_1 and f_2 converge to the solutions
of the differential equations:

$$\begin{cases} \dfrac{df_1(\tau)}{d\tau} = 1 - 2f_1(\tau) \\[2mm] \dfrac{df_2(\tau)}{d\tau} = 2(f_1(\tau) - 2f_2(\tau)) \end{cases}$$

The variance $v(\tau) = f_2(\tau) - f_1^2(\tau)$ is hence determined by

$$\frac{dv(\tau)}{d\tau} = 2f_1 - 4f_2 - 2f_1(1 - 2f_1) = -4v(\tau) \ ,$$

so the solution is:

$$\begin{cases} f_1(\tau) = \dfrac{1}{2} + c \cdot e^{-2\tau} \\[2mm] v(\tau) = v(o) \, e^{-4\tau}. \end{cases}$$

We hence see that if we start e.g. from an initial state where

$P_n = f^n(1-f)^{N-n} \binom{N}{n}$ then $f(o)$ is very close to f and $v(o) = 0(\frac{1}{N})$,

so it will remain very small, and $f(\tau)$ will be very close to $f_1(\tau)$ at any time τ if $f_1(o) = f$. Hence we see that with high precision $f(\tau)$ evolves completely deterministically according to the "B-equation"

$\dfrac{df}{d\tau} = 1 - 2f$ whose solution rapidly approaches the equilibrium value $f = \dfrac{1}{2}$.

In order to get a feeling for how rapidly the irreversible behaviour appears as $N \to \infty$ it is instructive to compare the average recurrence time for states near equilibrium $f = \dfrac{1}{2}$ and for states further away from $f = \dfrac{1}{2}$. It is well known that $t_f =$ average recurrence time $= P_{Nf}^{-1} =$ $= 2^N \binom{N}{Nf}^{-1}$. Hence near $f = \dfrac{1}{2}$ the local central limit theorem for $x = \sqrt{N} (f - \dfrac{1}{2})$ with variance $\dfrac{1}{4}$ tells us that

$$t_{1/2+x/\sqrt{N}} \approx \frac{\sqrt{2\pi}}{2} \sqrt{N} \, e^{2x^2} \ ,$$

and at the scale $\tau = \dfrac{t}{N}$:

$$\tau_{1/2+x/\sqrt{N}} \sim \frac{1}{\sqrt{N}} \, e^{2x^2} \ .$$

However when $f - 1/2$ is fixed and $N \to \infty$ it is easy to see using Stirlings

formula that $\dfrac{\tau_f}{\tau_{1/2}} \sim e^{N(h(\frac{1}{2}) - h(f))}$.

States inside the regime of "normal" fluctuations hence have a very short recurrence time, whereas states at finite distance from $1/2$ have a super-astronomical recurrence time. Take for example $f = 1/2 + 0.001$, $N = 10^{23}$, $\tau_{1/2} = 10^{-12}$ sec., then $\tau_f \sim 10^{-12} e^{10^{17}} \sim 10^{10^{17}}$, which can be compared to the age of the earth $\sim 5 \cdot 10^9$ years $\sim 10^{16}$ sec. So we see that even at the "cosmic scale" (Borel's echélle cosmique) the recurrence times are "infinite".

We will see that it is a general fact that the probability of a macroscopic fluctuation $\sim e^{-N(\text{entropy difference})}$ in a large system.

After a long time Boltzmanns description tells us that $f = 1/2$, but if one studies the system with a higher resolution one finds that small fluctuations around this value occur all the time. Their evolution in the normal regime can be studied by a suitable scaling. Put $f(\tau) = 1/2 + \dfrac{x(\tau)}{\sqrt{N}}$. Then for the Markov process $x(\tau)$ we have:

$$\left\langle x(\tau + \tfrac{1}{N}) - x(\tau) \,\big|\, x(\tau) = x \right\rangle = \frac{\sqrt{N}}{N}(1 - 2f) = -\frac{2x}{N}$$

$$\left\langle (x(\tau + \tfrac{1}{N}) - x(\tau))^2 \,\big|\, x(\tau) = x \right\rangle = N \left\langle (f(\tau + \tfrac{1}{N}) - f(\tau))^2 \,\big|\, f(\tau) = f \right\rangle =$$

$$= N\left[(1 - f)(f + \tfrac{1}{N})^2 + f(f - \tfrac{1}{N})^2 + f^2 - 2f(f + \tfrac{1}{N}(1 - 2f))\right] = \frac{1}{N} .$$

These infinitesimal moments are those which characterize a diffusion $dx(\tau) = -2x(\tau)d\tau + dw(\tau)$, and in this case it is not difficult to show that $x(\tau)$ converges to an Ornstein-Uhlenbeck diffusion with forward equation

$$\frac{\partial P}{\partial \tau} = \frac{1}{2}\frac{\partial^2 P}{\partial x^2} + \frac{\partial}{\partial x}(2xP) .$$

This is a Gaussian diffusion with stationary distribution $\dfrac{e^{-2x^2}}{\sqrt{2\pi}(\tfrac{1}{2})}$.

It is also reversible (like all one dimensional stationary diffusions). The process $x(\tau)$ can be described as the solution to the stochastic differential equation $\dfrac{dx}{d\tau} = -2x + \dfrac{dw}{d\tau}$, where $\dfrac{dw}{d\tau}$ is white noise. This equation is the B-equation for $f - 1/2$ with white noise added as a driving term.

We hence see that we have the following description of the dynamics when N is large:

If we start with $\frac{X(o)}{N} = f(o)$ fixed $\neq 1/2$ we see with high accuracy a deterministic motion towards equilibrium $\frac{X(t)}{N} = \frac{1}{2}$. When we have come to the normal regime $\frac{X(t)}{N} - \frac{1}{2} = \frac{x(\tau)}{\sqrt{N}}$ we can see fluctuations $x(\tau) = 0(1)$. Their dynamics is well described by a diffusion process which is reversible and is defined by the B-equation near equilibrium with white noise added in the right hand side. All the time we consider the system on the time scale $t = N \cdot \tau$. The relation between the stochastic and deterministic evolution is illustrated on the following diagram, which shows three outcomes of simulations of the Ehrenfest process with $N = 1000$ particles and the corresponding solutions of the B-equation. The standard deviation in equilibrium is $\sigma \approx 16$.

This picture is believed to be found in general even for complicated realistic systems even if one can not prove it completely, and we have seen how reversible and irreversible behaviour are related.

References: The Ehrenfest model and its approach to equilibrium is discussed by Kac in ref. 6.
Limit theorems for Markov jump processes converging to the solution of a differential equation are proved by Kurtz in:
Kurtz, T.G. Solutions of ordinary differential equations as limits of pure jump Markov processes. J. Appl. Prob. 7, 49-58 (1970) and
Limit theorems for sequences of jump Markov processes approximating ordinary differential processes. J. Appl. Prob. 8, 344-356 (1971).

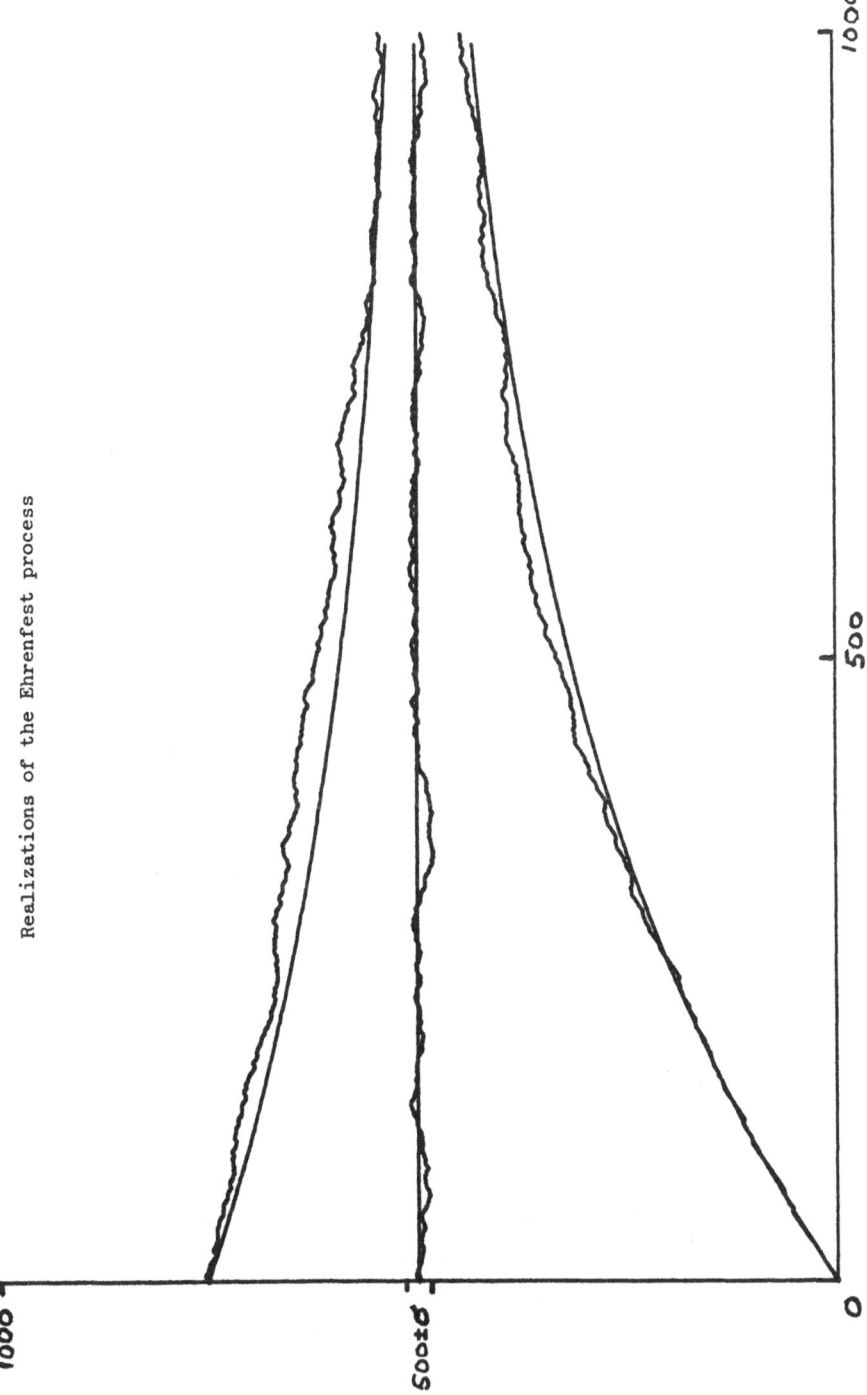

Realizations of the Ehrenfest process

2. Study of equilibrium distributions

2.1. The microcanonical distribution

In general it is difficult to compute averages in the m.can. distribution.
One exception is an ideal gas. Such a gas is characterised by the fact that
one neglects completely the contribution from $V(q)$ to $H(p,q)$ in the
calculation of all integrals. We hence have:

$$H(p,q) = \sum_1^N \frac{|p_i|^2}{2m} \quad, \quad \text{and if} \quad |\Lambda| = V = \text{volume:}$$

$$\Omega_\Lambda(E,N) = \int \frac{dpdq}{N!} = C_{3N}(2mE)^{\frac{3N}{2}} \frac{V^N}{N!}$$

$$\sum_1^N |p_i|^2 \leq 2mE$$
$$q_i \in \Lambda$$

with C_n = volume of unit sphere in $R^n = \frac{\pi^{n/2}}{(n/2)!}$. When we want to compute
the average of some function $f(x)$ we can do it as follows:

$$\langle f \rangle_{E,N,\Lambda} = \int_{\Gamma_E} f d\mu = \frac{1}{\Omega_\Lambda'(E,N)} \int_{H=E} \frac{f ds}{|\text{grad } H|N!} = \frac{1}{\Omega_\Lambda'(E,N)} \frac{d}{dE} \int_{H(x) \leq E} f(x) \frac{dx}{N!}$$

if $f(x)$ is defined also outside Γ_E. We shall always in the following regard
similar particles as indistinguishable, i.e. we shall only consider functions
$f(x)$ which are symmetric when such particles are permuted. Then the conven-
tion of introducing an $N!$ will turn out to be natural.

Let us study the description of a large system in the limit $E,N,V,\Lambda \to \infty$
such that $\frac{E}{V} \to e$, $\frac{N}{V} \to n$ with e,n finite. This is called the "thermodynamic
limit". The distribution of the state of the particles in any finite sub-
volume Λ_1 can then be computed by the following argument:

$$\Lambda = \Lambda_1 \cup \Lambda_2, \quad \text{and} \quad \Lambda_2 \to \infty \text{ too.}$$

If there are N_1 particles in Λ_1 the total energy is split into $H_1(x^{(1)}) +$
$+ H_2(x^{(2)})$, where $x^{(1)}$ and $x^{(2)}$ are the states in Λ_1 and Λ_2. If $f_1(x^{(1)}, N_1)$
is any symmetric function of the state of the particles in Λ_1 then its average
can be written as follows: For any $x = (x_1, x_2, \ldots, x_N)$ let $S(x)$ be the subset of $\{1,2,\ldots,N\}$ for which $x_i \in \Lambda_1$. x can then be split into $x^{(1)}(x) =$
$= (x_1^{(1)}, \ldots, x_{N_1}^{(1)})$ and $x^{(2)}(x) = (x_1^{(2)}, \ldots, x_{N_2}^{(2)})$, where the $x_n^{(i)}$ are

placed in the same order as in x. (E.g. if $x = (x_1,\ldots,x_5)$ $S(x) = \{2,3,5\}$ then $x^{(1)} = (x_2,x_3,x_5)$, $x^{(2)} = (x_1,x_4)$.)

Then:

$$\int_{H(x)\leq E} f_1(x^{(1)}(x), N_1(x)) \frac{dx}{N!} = \sum_{\substack{S \ H(x)\leq E \\ S(x)=S}} \int f_1(x^{(1)}(x), N_1(x)) \frac{dx}{N!} =$$

$$= \sum_{\substack{S \\ H_1(x^{(1)})+H_2(x^{(2)})\leq E}} \int f_1(x^{(1)}, |S|) \frac{dx^{(1)}dx^{(2)}}{N!} =$$

$$= \sum_{N_1} \binom{N}{N_1} \int \frac{1}{N!} \ f_1(x^{(1)}, N_1)dx^{(1)} \int_{H_2(x^{(2)})\leq E-H_1(x^{(1)})} dx^{(2)} =$$

$$= \sum_{N_1} \int f_1(x^{(1)}, N_1)\Omega_{\Lambda_2}(E-H_1(x^{(1)}), N-N_1) \frac{dx^{(1)}}{N_1!}, \quad \text{and}$$

$$\langle f_1 \rangle_{E,N,\Lambda} = \frac{1}{\Omega'_\Lambda(E,N)} \sum_{N_1} \int f_1(x^{(1)},N_1) \ \Omega'_{\Lambda_2}(E-H_1(x^{(1)}), N-N_1) \frac{dx^{(1)}}{N_1!}$$

We hence see that the probability of having N_1 particles in Λ_1 with state x_1 has probability density

$$\frac{\Omega'_{\Lambda_2}(E-H_1(x_1), N-N_1)}{\Omega'_\Lambda(E,N)} \ \frac{dx_1}{N_1!} \ .$$

In this argument we only used the fact that $H(x) = H_1(x^{(1)}) + H_2(x^{(2)})$ but not that H_1 and H_2 correspond to an ideal gas. Putting $f_1(x^{(1)},N_1) = 1$ iff $H_1(x^{(1)}) \leq E_1$, $N_1(x) = N_1$ we see that

$$P(H_1(x^{(1)}) \leq E_1, \ N_1(x) = N_1) =$$

$$= \frac{1}{\Omega'_\Lambda(E,N)} \int_{H_1(x^{(1)})\leq E_1} \Omega'_{\Lambda_2}(E-H_1(x^{(1)}),N-N_1) \frac{dx^{(1)}}{N_1!} =$$

$$= \frac{1}{\Omega'_\Lambda(E,N)} \int_{-\infty}^{E_1} \Omega'_{\Lambda_1}(H_1,N_1)\Omega'_{\Lambda_2}(E-H_1,N-N_2)dH_1 \ ,$$

so the simultaneous density of E_1 and N_1 is

$$\frac{\Omega'_{\Lambda_1}(E_1,N_1)\ \Omega'_{\Lambda_2}(E-E_1,N-N_1)}{\Omega'_\Lambda(E,N)}\ dE_1\ .$$

From this result we see that $\Omega'_\Lambda(E,N)$ is composed by convolution, because the total integral is one:

$$\Omega'_\Lambda(E,N) = \sum_{N_1} \int \Omega'_{\Lambda_1}(E_1,N_1)\ \Omega'_{\Lambda_2}(E-E_1,N-N_1)dE_1\ .$$

Let us sum up these results:

Lemma 1. Consider a system in Λ with a m.can. distribution defined by E,N. Then if $\Lambda = \Lambda_1 \cup \Lambda_2$ with $\Lambda_1 \cap \Lambda_2 = \phi$ and if there is no interaction between particles in Λ_1 and Λ_2 then N_1 and the state in Λ_1 has probability density

$$\frac{\Omega'_{\Lambda_2}(E-H_1(x_1),N-N_1)}{\Omega'_\Lambda(E,N)}\ \frac{dx_1}{N_1!}$$

and hence E_1,N_1 have density

$$\frac{\Omega'_{\Lambda_1}(E_1,N_1)\ \Omega'_{\Lambda_2}(E-E_1,N-N_1)}{\Omega'_\Lambda(E,N)}\ dE_1$$

and

$$\Omega'_\Lambda(E,N) = \sum_{N_1} \int \Omega'_{\Lambda_1}(E_1,N_1)\ \Omega'_{\Lambda_2}(E-E_1,N-N_1)\ dE_1\ .$$

Returning now to the case of an ideal gas we have in this case

$$\Omega'_\Lambda(E,N) = \frac{\pi^{\frac{3N}{2}}}{(\frac{3N}{2})!}\ (\frac{3N}{2})\ (2m)^{\frac{3N}{2}}\ E^{\frac{3N}{2}-1}\ \frac{V^N}{N!}\ ,$$

so in the limit $N \to \infty$, $\frac{E}{N} \to \epsilon$, $\frac{N}{V} \to n$

$E_1,V_1,N_1 = $ const.

$$\frac{\Omega'_{\Lambda_2}(E_2,N_2)}{\Omega'_\Lambda(E,N)}\ \frac{dx_1}{N_1!} =$$

$$\frac{dx_1}{N_1!}\ (2m\pi)^{-\frac{3N_1}{2}}\ \frac{(\frac{3N}{2})!}{(\frac{3N_2}{2})!}\ \frac{E_2^{\frac{3N_2}{2}-1}}{E^{\frac{3N}{2}-1}}\ \frac{V_2^{N_2}}{V^N}\ \frac{N!}{N_2!}\ \approx$$

$$\approx \frac{dx_1}{V_1^{N_1}} \left(\frac{V_1}{V}\right)^{N_1} \left(\frac{V_2}{V}\right)^{N_2} \binom{N}{N_1} (2m\pi)^{\frac{-3N_1}{2}} \frac{\left(\frac{3N}{2e}\right)^{\frac{3N}{2}}}{\left(\frac{3N_2}{2e}\right)^{\frac{3N_2}{2}}} \frac{E_2^{\frac{3N_2}{2}}}{E^{\frac{3N}{2}}}$$

$$\approx \frac{dx_1}{V_1^{N_1}} \frac{(nV_1)^{N_1}}{N_1!} e^{-nV_1} (2m\pi)^{-\frac{3N_1}{2}} \left(\frac{3}{2e}\right)^{\frac{3N_1}{2}} \left(\frac{N}{N_2}\right)^{\frac{3N}{2}} \left(\frac{E_2}{E}\right)^{\frac{3N}{2}} \left(\frac{N_2}{E_2}\right)^{\frac{3N_1}{2}}$$

The last factors tend to:

$$\left(1 - \frac{N_1}{N}\right)^{\frac{-3N}{2}} \longrightarrow e^{\frac{3N_1}{2}} \quad \text{and}$$

$$\left(1 - \frac{H_1}{E}\right)^{\frac{3N}{2}} \longrightarrow e^{\frac{-3H_1}{2\varepsilon}}$$

$$\left(\frac{N_2}{E_2}\right)^{\frac{3N_1}{2}} \longrightarrow \varepsilon^{\frac{-3N_1}{2}},$$

so the probability density for N_1 and x_1 converges to

$$\frac{(nV_1)^{N_1}}{N_1!} e^{-nV_1} \left(\frac{4m\pi\varepsilon}{3}\right)^{\frac{-3N_1}{2}} e^{\frac{-3H_1(x_1)}{2\varepsilon}} \frac{dx_1}{V_1^{N_1}} =$$

$$= \frac{(nV_1)^{N_1}}{N_1!} e^{-nV_1} \left(\frac{4m\pi\varepsilon}{3}\right)^{\frac{-3N_1}{2}} e^{\frac{-3}{4m\varepsilon} \sum_1^{N_1} |p_i|^2} \frac{dp\ dq}{V_1^{N_1}} .$$

We hence see that the probabilistic description of an ideal gas in equilibrium is the following: The positions in space form a Poisson process with density n and all the momenta are independent and have a normal (Maxwellian) distribution with variance $\frac{2m\varepsilon}{3}$, so the average kinetic energy of a particle is $\left\langle \frac{|p_i|^2}{2m} \right\rangle = \varepsilon$.

In this case we see that the following law holds: The distribution of the

state of a small subsystem in a large system with finite densities of particles and energy has a probability density proportional to

$$e^{-\beta H_N(x)+\beta\mu N} \frac{dx}{N!} \; ,$$

where β and $\beta\mu$ are parameters. This is a very general law which was disco-
vered by Boltzmann. (The exponential factor is often called the B-factor.) It
is easy to verify also in the situation where we have an arbitrary system in Λ_1
with a fixed no. of particles in contact with an ideal gas in Λ_2, a heat
bath. The two parts can exchange energy, but the interaction energy can be
neglected. The particles in Λ_1 and Λ_2 need not to be of the same type. Then

$$\Omega_\Lambda(E,N_1,N_2) = \int_{H_1(x_1)+H_2(x_2)\leq E} \frac{dx_1}{N_1!} \frac{dx_2}{N_2!} = \int \frac{dx_1}{N_1!} \Omega_{\Lambda_2}(E-H_1(x_1),N_2) =$$

$$= \int \Omega'_{\Lambda_1}(E_1,N_1)\, \Omega_{\Lambda_2}(E-E_1)dE_1$$

(The particles should not be permuted between Λ_1 and Λ_2.)

We see as in lemma 1 that the probability density for x_1 is

$$\frac{\Omega'_{\Lambda_2}(E-H_1(x_1),N_2)}{\Omega'_\Lambda(E,N_1,N_2)} \frac{dx_1}{N_1!} \; .$$

Since

$$\Omega'_{\Lambda_2}(E_2,N_2) = K(N_2)E_2^{\frac{3N_2}{2}-1} \; , \quad \text{we see that the density is proportional to}$$

$$(E-H_1(x_1))^{\frac{3N_2}{2}-1} = (\text{const})\left(1 - \frac{H_1(x_1)}{E}\right)^{\frac{3N_2}{2}-1} \longrightarrow$$

$$\longrightarrow (\text{const})\, e^{\frac{-3H_1(x_1)}{2\varepsilon}} \; , \quad \text{and it hence converges to}$$

$$\frac{e^{-\beta H_1(x_1)}}{\int \Omega'_{\Lambda_1}(E_1,N_1)\, e^{-\beta E_1}dE_1} \frac{dx_1}{N_1!} \qquad \beta = \frac{3}{2\varepsilon}$$

This distribution is called the canonical and is more important in practice than
the m. can. one since very often one is interested in precisely this situation
where the system is in contact with a "heath bath". We shall see that in general

$H_1(x_1)$ has a distribution very sharply concentrated around its average when N_1 is large, so one gets the same results concerning most averages as in the m.can. "ensemble" where $H_1(x_1)$ is fixed to this value, because the conditional distribution of x_1 given that $H_1(x_1) = E_1$ is the m.can. distribution. (The B-factor is constant on each energy shell.) The exponential form means that if two systems do not interact: $H(x_1,x_2) = H_1(x_1) + H_2(x_2)$ then they are also independent:

$$e^{-\beta H(x_1,x_2)} = e^{-\beta B_1(x_1)} e^{-\beta H_2(x_2)}, \quad \text{and this is a considerable simplification.}$$

(Compare this to coin tossing: $x = (x_1,\ldots,x_N)$

$$x_i = \begin{cases} 0 \\ 1 \end{cases} \quad \text{Put} \quad H(x) = \sum_1^N x_i . \quad \text{Then}$$

$\Omega'(E,N) = $ (no. of x s.t. $H(x) = E) = \dbinom{N}{E}$. The distribution of $x^{(1)} = (x_1,\ldots,x_n)$, with $H_1(x^{(1)}) = \sum_1^n x_i$ is given by

$$P(x^{(1)}) = \frac{\Omega_2'(E-H_1(x^{(1)}))}{\Omega'(E)} = \frac{\dbinom{N-n}{E-H_1}}{\dbinom{N}{E}} = \frac{(N-n)! \; E! \; (N-E)!}{N! \; (E-H_1)! \; (N-n-E+H_1)!}$$

$$\approx \frac{E^{H_1}(N-E)^{n-H_1}}{N^n} \longrightarrow \varepsilon^{H_1}(1-\varepsilon)^{n-H_1} = \text{Bernoulli distribution as } N \to \infty$$

$\frac{E}{N} \to \varepsilon$. This distribution is much simpler to handle than the m.can. one.)

A digression concerning the definition of the uniform measure on an infinite dimensional sphere

The argument above has shown that the uniform measure on the sphere

$$S_N: \sum_1^N x_i^2 = N \text{ in } R^N$$

converges to the infinite product measure of independent Gaussians on R^∞ in the sense that for each n the density of $(x_1,\ldots,x_n)$ converges to

$(2\pi)^{-\frac{n}{2}} e^{-\frac{1}{2}\sum_1^n x_i^2}$. This measure should hence in some sense be considered as the uniform measure on the infinite dimensional sphere S_∞ "$\sum_1^\infty x_i^2 = \infty$".

This measure can also be identified with Wienermeasure on

$C(0,\infty)$ via the following map: Take an orthonormal basis $\{\phi_n\}$ for $L^2(0,\infty)$ and put $x_n = \int_0^\infty \phi_n(t)W(dt)$. (Itô integral.) Then we have $E(x_n,x_m) = (\phi_n,\phi_m) = \delta_{n,m}$ and $\{x_n\}$ are independent Gaussian, i.e. have the "spherical" distribution. This point of view is fruitful because it shows that several concepts defined for S_N can be generalized to S_∞ also:

The measure is invariant with respect to orthogonal transformations $x_n \to \Sigma_m O_{nm} x_m$. For $L^2(S_N)$ there is a natural orthogonal base system, the so called spherical polynomials. They are eigenfunctions of the Laplacian, and can be defined by

$$P_m(x) = D^m |x|^{2-N} = \frac{\partial^{m_1}}{\partial x_1^{m_1}} \frac{\partial^{m_2}}{\partial x_2^{m_2}} \cdots |x|^{2-N}$$

$(m = (m_1,\ldots,m_N) = \text{multiindex}, \; |m| = \sum_1^N m_i)$

These have simple limits as $N \to \infty$:

$$\lim_{N\to\infty} N^{\frac{N}{2}-1} P_m(x) = (-1)^{|m|} H_m(x)$$

$$H_m(x) = h_{m_1}(x_1) \cdot h_{m_2}(x_2) \cdots$$

with $h_n(x) = $ Hermitepolynomials defined by

$$h_n(x) = (-1)^n e^{\frac{x^2}{2}} D^n(e^{-\frac{x^2}{2}})$$

Proof: Choose M so that $m_i = 0$ for $i \geq M$. Since $\sum_1^N x_i^2 = N$ we then have

$$N^{\frac{N}{2}-1} P_m(x) = N^{\frac{N}{2}-1} D^m \left[x_1^2 + \cdots + x_M^2 + \cdots x_N^2 \right]^{1-\frac{N}{2}} =$$

$$= N^{\frac{N}{2}-1} \left[x_{M+1}^2 + \cdots + x_N^2 \right]^{1-\frac{N}{2}} D^m \left[1 + \frac{x_1^2 + \cdots + x_M^2}{x_{M+1}^2 + \cdots + x_N^2} \right]^{1-\frac{N}{2}} =$$

$$= \left[1 - \frac{x_1^2 + \cdots + x_M^2}{N} \right]^{1-\frac{N}{2}} D^m \left[1 + \frac{x_1^2 + \cdots + x_M^2}{N + O(N)} \right]^{1-\frac{N}{2}} \to$$

$$\to e^{\frac{1}{2}(x_1^2 + \cdots + x_M^2)} D^m \; e^{-\frac{1}{2}(x_1^2 + \cdots + x_M^2)} = (-1)^{|m|} H_m(x)$$

It is now true that also in the limit $N \to \infty$ $\{H_m\}$ forms an orthogonal basis for $L^2(S_\infty) = L^2$ (Wienerspace) in the following way: Define the random variables H_m by

$$H_m = h_{m_1} \left(\int \phi_1(t)W(dt) \right) \cdot h_{m_2} \left(\int \phi_2(t)W(dt) \right) \cdots$$

It is then true that every random variable $X \in L^2(\text{Wiener})$ can be expanded in this basis: $X = \sum_m H_m \cdot E(X \cdot H_m)$. This basis was introduced by Wiener, and he called $\{H_m\}$ "polynomial chaoses" of orders $|m|$. One can also define the limit of the Laplace operator Δ_N on S_N, and it becomes

$$\Delta = \sum_i \frac{\partial^2}{\partial x_i^2} - x_i \frac{\partial}{\partial x_i}$$

for nice functions $\phi = \phi(x_1, \ldots, x_n)$. As on S_N the H_m are eigenfunctions of Δ on S_∞. This construction of S_∞ is of interest in the constructive theory of quantum fields. All these relations are explained in detail in the interesting article "Geometry of Differential Space" by H.P. McKean, Ann. Prob. 1 (1973), 197.

2.2. The canonical distribution

We shall now study a system with a can. probability law:

$$f(x) \frac{dx}{N!} = \frac{e^{-\beta H(x)}}{Z_\Lambda(\beta, N)} \frac{dx}{N!},$$

$$Z_\Lambda(\beta, N) = \int e^{-\beta H(x)} \frac{dx}{N!},$$

and give the physical meaning of β and the "equation of state" of the system. For an ideal gas we hence have

$$f(x) = \frac{e^{-\beta \sum_1^N \frac{|p_i|^2}{2m}}}{Z} \quad \text{for } q_i \in \Lambda,$$

so

$$Z_\Lambda(\beta, N) = \frac{V^N}{N!} \frac{(2\pi m)^{\frac{3N}{2}}}{\beta}.$$

For a gas with interaction we have

$$f(x) = \frac{e^{-\beta \sum_1^N \frac{|p_i|^2}{2m}} \; e^{-\beta \sum_{i<j} V(q_i - q_j)}}{Z},$$

so we see that still all $\{p_i\}$ are independent gaussian and independent of $\{q_i\}$. The contribution to Z from the p integration can easily be computed like for an ideal gas, so we have

$$Z_\Lambda(\beta, N) = \left(\frac{2\pi m}{\beta}\right)^{\frac{3N}{2}} Q_\Lambda(\beta, N) \quad \text{with}$$

$$Q_\Lambda(\beta,N) = \int e^{-\beta \sum_{i<j} V(q_i-q_j)} \frac{dq}{N!}$$

Q is often called the "configuration integral". For an arbitrary gas it is still true that $\langle |p_i|^2 \rangle = \frac{3m}{\beta}$, or $\frac{\langle |p_i|^2 \rangle}{2m} = \langle E_{kin,i} \rangle = \frac{3}{2\beta}$. The average kinetic energy of a particle is $\frac{3}{2\beta}$. This is still true if the masses are different. It is also true in the more general case when H is of the form

$$H(p,q) = \frac{1}{2} \sum_{ij=1}^{f} p_i\, p_j\, a_{ij}(q) + V(q) , \quad \text{because then}$$

$$\int p_i \frac{\partial H}{\partial p_j} e^{-\beta H} dp = \int \frac{p_i}{-\beta} \frac{\partial}{\partial p_j} (e^{-\beta H}) dp =$$

$$= I \frac{p_i}{-\beta} e^{-\beta H} + \frac{1}{\beta} \int \delta_{ij} e^{-\beta H} dp = \frac{\delta_{ij}}{\beta} \int e^{-\beta H} dp$$

Hence $\langle p_i \frac{\partial H}{\partial p_j} \rangle = \frac{\delta_{ij}}{\beta}$, so that

$$\langle E_{kin} \rangle = \langle \frac{1}{2} \sum_{ij} p_i\, p_j\, a_{ij}(q) \rangle = \langle \frac{1}{2} \sum_i p_i \frac{\partial H}{\partial p_i} \rangle = \frac{f}{2\beta} .$$

This is the so called equipartition law:

Lemma 2. If the kinetic energy is a quadratic form in p_i with f "degrees of freedom", then $\frac{\langle E_{kin} \rangle}{f} = \frac{1}{2\beta}$.

Example. Consider a system of harmonic occillators:

$$H(p,q) = \frac{1}{2} \sum_{1}^{N} \frac{|p_i|^2}{2m} + \frac{1}{2} \sum_{1}^{N} q_i^T\, V q_i$$

This is a model for small occillations of the molecules of a solid.
q_i = the displacement of molecule i from its equilibrium position,
and the potential energy is a quadratic function of the q_i. The equipartition law for both p and q then says that

$$\langle H \rangle = \frac{3N}{2\beta} + \frac{3N}{2\beta} = \frac{3N}{\beta} .$$

This law describes the total average energy as a function of β (temperature) and says what is the specific heat. (Dulong-Petit's law)

The equation of state and the identification $\beta = \frac{1}{kT}$

The equation of state of a system gives the relation between p,V,N,T, pressure, volume, no. of particles and temperature in equilibrium. To determine it we must calculate the pressure and the temperature.

The pressure p is the force per unit area of the container which one has to supply to balance the effect of the collisions of the particles with the wall:

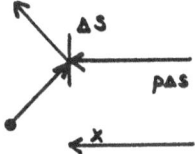

In a collision the momentum of the particle is changed from (p_x, p_y, p_z) to $(-p_x, p_y, p_z)$. The change is hence $\Delta p = 2p_x$ in the direction of the normal pointing into Λ. The average of $\Sigma\Delta p$ for all collisions with $\partial\Lambda$ during a short time interval Δt gives the momentum which has to be supplied to the particles from $\partial\Lambda$, and this is (total force) $\times\Delta t$. Hence p is given by

$p\cdot|\partial\Lambda|\cdot\Delta t = \langle\Sigma\Delta p\rangle$ in the limit $\Delta t \to 0$.
    collisions
    to $\partial\Lambda$ in Δt

To calculate $\langle\Sigma\Delta p\rangle$ consider a particle very near to $\partial\Lambda$ at distance x from

it. It will see $\partial\Lambda$ as locally flat. Consider a short Δt so small that we can neglect the probability that the particle will collide with another particle before hitting $\partial\Lambda$. The particle will hit $\partial\Lambda$ in the next interval Δt if $-v_x \geq \frac{x}{\Delta t}$, and then it will have $\Delta p = 2p_x = 2m|v_x|$. Hence the average of Δp for such a particle is

$$\int_{\frac{mx}{\Delta t}}^{\infty} \frac{e^{-\frac{\beta p^2}{2m}}}{\sqrt{\frac{2\pi m}{\beta}}} 2p \, dp = \sqrt{\frac{\beta}{2\pi m}} \frac{2m}{\beta} e^{-\frac{\beta}{2m}(\frac{mx}{\Delta t})^2} = 2\sqrt{\frac{m}{2\pi\beta}} e^{-\frac{\beta m}{2}(\frac{x}{\Delta t})^2} .$$

Hence the total average is:

$$\langle\Sigma\Delta p\rangle = \int_\Lambda n(q) 2\sqrt{\frac{m}{2\pi\beta}} e^{-\frac{\beta m}{2}(\frac{x(q)}{\Delta t})^2} dq$$

where $n(q)$ is the particle density at q and $x(q)$ the distance from q to $\partial\Lambda$. Now, as $\Delta t \to 0$ the integrand will have a sharp peak at $\partial\Lambda$, so if $n(q)$ is nice only its value near $\partial\Lambda$ will be filtered out:

Use coordinates as indicated: $dq = dx\,ds$,

σ^2 the variance of the gaussian is $\dfrac{\Delta t^2}{\beta m}$.

Hence as $\Delta t \to 0$ we have

$$\langle \Sigma \Delta p \rangle = \int_{\partial\Lambda} ds \int_0^\infty n(q) 2\frac{\Delta t}{\beta} \frac{1}{\sqrt{2\pi}\sigma} e^{-\frac{x^2}{2\sigma^2}} dx$$

$$\approx \Delta t \int_{\partial\Lambda} ds \frac{n(q)}{\beta},$$

and we finally conclude that

$$p = \frac{1}{|\partial\Lambda|} \int_{\partial\Lambda} \frac{n(q)}{\beta} ds.$$

This expression for p can be directly related to $\log Z$. Consider namely a small change in Λ and the corresponding change in $\log Z$:

$$\Delta \log Z = \Delta \log Q = (N!Q)^{-1} \left[\int_{\substack{\text{all } q_i \text{ in} \\ \Lambda \cup \Delta\Lambda}} e^{-\beta V(q)} dq - \int_{\text{all } q_i \text{ in } \Lambda} e^{-\beta V(q)} dq \right]$$

$$= (N!Q)^{-1} \left[N \int_{\Delta\Lambda} dq_1 \int_{\Lambda^{N-1}} e^{-\beta V(q)} dq_2 \dots dq_N + O(\Delta V)^2 \right]$$

(The remainder is the integral with more than one particle in $\Delta\Lambda$.) But

$$\frac{N}{Q} \int_{\Lambda^{N-1}} e^{-\beta V(q)} \frac{dq_2 \dots dq_N}{N!}$$

is the probability density for finding a particle at $q_1 = n(q_1)$.

Hence

$$\Delta \log Z = \int_{\Delta\Lambda} n(q_1)dq_1 + O(\Delta V)^2 =$$

$$= \Delta V \cdot \frac{1}{\left|\frac{\partial\Lambda}{\partial\Lambda}\right|} \int_{\partial\Lambda} n(q)dq + O(\Delta V)^2,$$

so we see that

$$p = \lim_{\Delta V \to 0} \frac{1}{\beta} \frac{\Delta \log Z_\Lambda(\beta,N)}{\Delta V} .$$

For a large system it will turn out that $\log Z_\Lambda(\beta,N)$ is very insensitive to the shape of Λ, so one usually writes

$$p = \frac{1}{\beta} \frac{\partial \log Z_\Lambda(\beta,N)}{\partial V}$$

for a large system. An alternative derivation of the formula for the pressure: Consider the system together with a small "manometer" consisting of a movable piston with a spring in the wall of the container.The position

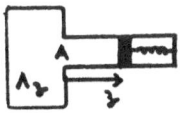

and momentum of the piston are also included as extra coordinates, and the total volume depends on z. Let the energy of the spring be $F(z)$. Then the total B-factor is

$$e^{-\beta H(p,q) - \beta F(z) - \beta M\frac{\dot{z}^2}{2}} \qquad q_i \in \Lambda_z$$

(M = the mass of the piston.)

The average of the force on the piston is measured by the manometer as $p \cdot A$ and can be expressed as follows:

$$\langle F'(z) \rangle = Z^{-1} \int dz\, F'(z)\, e^{-\beta F(z)} \int_{q \in \Lambda_z^N} e^{-\beta H(p,q)} \frac{dpdq}{N!}$$

$$= Z^{-1} \int dz\, F'(z)\, e^{-\beta F(z)}\, Z(z) =$$

$$= Z^{-1} \int Z(z)(-\frac{1}{\beta}) \frac{\partial}{\partial z} e^{-\beta F(z)} dz =$$

$$= Z^{-1}\beta^{-1} \int \frac{\partial Z(z)}{\partial z} e^{-\beta F(z)} dz =$$

$$= \beta^{-1} Z^{-1} \int \frac{\partial \log Z(z)}{\partial z} Z(z) \ e^{-\beta F(z)} dz = \beta^{-1} \left\langle \frac{\partial \log Z(z)}{\partial z} \right\rangle$$

(We have neglected the effect of $\frac{M\dot{z}^2}{2}$ which is completely cancelled.)

For a macroscopically large manometer the fluctuations of z ought to be small, and then we have with high accuracy

$$p = \frac{\langle F'(z) \rangle}{A} = \frac{1}{\beta A} \frac{\partial \log Z}{\partial z} = \frac{1}{\beta} \frac{\partial \log Z}{\partial V} ,$$

the same formula as before.

For the ideal gas we have $\log Z = const + N \log V$, and hence $p = \frac{N}{\beta V}$, or $pV = \frac{N}{\beta}$. From physics in school we remember that the ideal gas law (Boyle) is $pV = kNT$, where T is temperature (on the absolute scale). We have hence derived this law if we identify $\beta = \frac{1}{kT}$. This relation ought then to be valid in general for an arbitrary system, because the canonical distribution describes the state of the system in thermal equilibrium with a heat bath consistning of an ideal gas e.g. If the whole system is described by a canonical distribution then β as we just said is $\frac{1}{kT}$, where T is the temperature of the bath. But then it ought to be also the temperature of the system because two systems are in thermal equilibrium iff their temperatures are equal. (We will elaborate this argument later.)

For a non ideal gas the equation of state is hence defined by

$$p = \frac{1}{\beta} \frac{\partial \log Z_\Lambda(\beta,N)}{\partial V} = \frac{1}{\beta} \frac{\partial \log Q_\Lambda(\beta,N)}{\partial V}$$

$$\beta = \frac{1}{kT} .$$

Boyle's law can be written $\frac{pV}{kT} = N$, and then it says that for ideal gases with the same p,V,T N is also the same. This is Avogadro's law. Boyle's law can also easily be extended to a gas located in an external field with potential $U(q)$, e.g. the atmosphere in the field of gravity. The canonical probability density is then

$$f(p,q) = Z_\Lambda^{-1}(\beta,N) \ e^{-\beta \sum_1^N \frac{|p_i|^2}{2m} \ -\beta \sum_1^N U(q_i)} \frac{1}{N!}$$

$$Z_\Lambda(\beta,N) = (\frac{2\pi m}{\beta})^{\frac{3N}{2}} \frac{\left(\int_\Lambda e^{-\beta U(q)} dq \right)^N}{N!}$$

The probability density $n(q)dq$ of finding a particle at $q \in \Lambda$ is then given by

$$n(q) = \frac{Ne^{-\beta U(q)} \displaystyle\int_\Lambda e^{-\sum\limits_{2}^{N} U(q_i)} \dfrac{dq_2 \cdots dq_N}{N!}}{\dfrac{\left(\displaystyle\int_\Lambda e^{-\beta U(q)} dq\right)^N}{N!}}$$

$$n(q) = \frac{Ne^{-\beta U(q)}}{\displaystyle\int_\Lambda e^{-\beta U(q)} dq} \quad , \text{ so the pressure is } p(q) = \frac{n(q)}{\beta} \ .$$

Example Λ is a vertical cylinder in the field of gravity:
$U(z) = m \cdot g \cdot z$

$$\int_\Lambda e^{-\beta U(q)} dq = A \int_0^\infty e^{-\beta mgz} dz = \frac{A}{\beta mg}$$

$$n(q) = \beta mg \frac{N}{A} e^{-\beta mgz} \qquad q = (x,y,z)$$

$$p(q) = mg \frac{N}{A} e^{-\beta mgz}$$

This is the so called barometric formula.

Fluctuations of the energy in the canonical distribution

From the formula for Z_Λ :

$$Z_\Lambda(\beta, N) = \int e^{-\beta H(x)} \frac{dx}{N!} \qquad \text{we see that}$$

$$\frac{\partial^k Z_\Lambda}{\partial \beta^k} = \int (-H(x))^k e^{-\beta H(x)} \frac{dx}{N!} \qquad k = 1,2,\ldots$$

and hence

$$\langle H(x) \rangle = - Z_\Lambda^{-1} \frac{\partial Z_\Lambda}{\partial \beta} = - \frac{\partial \log Z_\Lambda}{\partial \beta}$$

$$\text{Var}(H(x)) = Z_\Lambda^{-1} \frac{\partial^2 Z_\Lambda}{\partial \beta^2} - (\frac{\partial \log Z_\Lambda}{\partial \beta})^2 = \frac{\partial^2 \log Z_\Lambda}{\partial \beta^2} = - \frac{\partial \langle H(x) \rangle}{\partial \beta}$$

$\langle H \rangle$ is the total average energy and $\frac{\partial \langle H \rangle}{\partial kT} = - \frac{1}{\beta^2} \frac{\partial \langle H \rangle}{\partial \beta}$ is called the specific heat C_v (at constant volume). Hence we have $\text{Var}(H) = \beta^2 C_v$.

For a large system it is true under quite general assumptions that $\log Z_\Lambda$, $\langle H \rangle$ and C_v are extensive, i.e. they grow essentially proportionally to V as $\Lambda, V, N \to \infty$ $\frac{N}{V} \to n$, as we will see later. (This is consistent with our every day experience, it takes twice as much energy to heat 2 l of water one degree than to heat 1 l e.g.).

For such a system with a finite energy and specific heat per unit volume we hence have $\langle \frac{H}{V} \rangle$ finite and $\text{Var}(\frac{H}{V}) = \frac{1}{V}(\frac{C_v}{V}) = O(\frac{1}{V})$, so $\frac{H}{V}$ has a very sharply concentrated distribution as $V \to \infty$, and one ought to be able to conclude that the canonical law gives asymptotically the same results as a microcanonical one with $E = \langle H \rangle$. We will verify this in the following.

2.3. The grand canonical distribution

We have seen that the can. distribution describes a system in equilibrium which can exchange energy but not particles with its surroundings. Another situation which is often encountered is that of an open subsystem, which can also exchange particles with the surroundings. For example when one considers the state of a small volume of a gas or more complicated chemical system and wants to study how the density, pressure, composition etc. varies. In a homogenous system one expects to get the same values of such variables as in the whole system, and as for the can. ensemble one can hope that it is simpler to relax the constraints as much as possible.

Let us hence as in Lemma 1 consider a subsystem Λ_1, and for simplicity let us asume that the interaction is of finite range R, so that $H(x_1, x_2) = H(x_1) + H(x_2)$ if x_1 and x_2 are two configurations a distance $\geq$ R apart. In lemma 1 we neglected the interaction between Λ_1 and the surroundings Λ_2, but now let us consider the distribution of the state x_1 in Λ_1 given that the state is fixed to x_Δ in a boundary layer Δ of width R surrounding

$\Lambda_1, \Lambda_2 = \Lambda \setminus (\Lambda_1 \cup \Delta):$

Then if x_2 is a configuration in Λ_2 there is only
interaction between x_1 and x_Δ, x_2 and x_Δ, but not
between x_1 and x_2, so $H(x_1, x_2) = H_1(x_1) + H_2(x_2)$,
where the interaction energy with x_Δ is included in H_1 and H_2 respectively
(although x_Δ is not explicitly shown in the notation.)

Then the argument of Lemma 1 shows that the conditional density of x_1 and
N_1 given x_Δ in the m. can. ensemble is given by:

$$f_{\Lambda_1}(x_1, N_1 | x_\Delta) \frac{dx_1}{N_1!} = \frac{\Omega'_{\Lambda_2}(E - H_1(x_1), N - N_1)}{\Omega'_{\Lambda}(E, N)} \frac{dx_1}{N_1!}$$

with

$$\Omega'_\Lambda(E, N) = \sum_{N_1} \int \Omega'_{\Lambda_1}(E_1, N_1) \Omega'_{\Lambda_2}(E - E_1, N - N_1) \, dE_1$$

Now, let us assume that the following limit exists as $\Lambda_2 \to \infty$ $\frac{E}{V} \to e, \frac{N}{V} \to n$:

$$\lim \frac{\Omega'_{\Lambda_2}(E - H_1, N - N_1)}{\Omega'_{\Lambda_2}(E, N)} = \omega(H_1, N_1), \text{ and that}$$

$$\lim \frac{\Omega'_\Lambda(E, N)}{\Omega'_{\Lambda_2}(E, N)} = \lim \sum_{N_1} \int \Omega'_{\Lambda_1}(E_1, N_1) \frac{\Omega'_{\Lambda_2}(E - E_1, N - N_1)}{\Omega'_{\Lambda_2}(E, N)} \, dE_1 =$$

$$= \sum_{N_1} \int \Omega'_{\Lambda_1}(E_1, N_1) \, \omega(E_1, N_1) \, dE_1$$

Then it is easy to see that $\omega(H_1, N_1)$ actually has to be of the exponential
form encountered for an ideal gas:

$$\omega(E_1, N_1) = e^{-\beta E_1 + \beta \mu N_1}$$

for some constants $-\beta, \beta\mu$. In fact

$$\omega(E' + E'', N' + N'') = \lim \frac{\Omega'_{\Lambda_2}(E - E' - E'', N - N' - N'')}{\Omega'_{\Lambda_2}(E, N)}$$

$$= \lim \frac{\Omega'_{\Lambda_2}(E - E' - E'', N - N' - N'')}{\Omega'_{\Lambda_2}(E - E', N - N')} \frac{\Omega'_{\Lambda_2}(E - E', N - N')}{\Omega'_{\Lambda_2}(E, N)} =$$

$= \omega(E'',N'')\cdot\omega(E',N')$, because $\dfrac{E - E'}{V} \to e$, $\dfrac{N - N'}{V} \to n$ also, and any Baire function satisfying this equation has to be an exponential. The limiting density is hence

$$f_{\Lambda_1}(x_1,N_1|x_\Delta)\,\frac{dx_1}{N_1!} = \frac{e^{-\beta H_1(x_1) + \beta\mu N_1}}{G_{\Lambda_1}(\beta,\mu|x_\Delta)}\,\frac{dx_1}{N_1!} \qquad \text{with}$$

$$G_{\Lambda_1}(\beta,\mu|x_\Delta) = \sum_{N_1}\int e^{-\beta H_1(x_1)+\beta\mu N_1}\,\frac{dx_1}{N_1!} = \sum_{N_1} e^{\beta\mu N_1} Z_{\Lambda_1}(\beta,N_1|x_\Delta)$$

This density is called the grand canonical density, and μ is usually called the chemical potential, and β is the inverse temperature. The density of N_1 is hence

$$\frac{e^{\beta\mu N_1} Z_{\Lambda_1}(\beta,N_1|x_\Delta)}{G_{\Lambda_1}(\beta,\mu|x_\Delta)}\quad,$$

and the conditional density of x_1 given N_1 and x_Δ is the canonical one with x_Δ fixed:

$$\frac{e^{-\beta H_1(x_1)}}{Z_{\Lambda_1}(\beta,N_1|x_\Delta)}\,\frac{dx_1}{N_1!}\quad.$$

Actually this derivation only shows that to each Λ_1 there is a β and a μ defining the g.can. density, but a priori different regions could have different β and μ . This is however not the case, because if we have two regions Λ_1 Λ_2 with $\Lambda_1 \cup \Lambda_1 \subset \Lambda_2$ and g.can. densities defined by $\beta_1,\mu_1,\beta_2,\mu_2$ respectively then the g.can. density in Λ_1 has to be equal to the one induced by the g.can. density in Λ_2, and the latter is given by the following calculation:

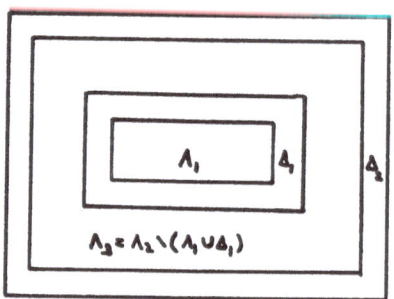

$(\beta = \beta_2,\ \mu = \mu_2,\ \Delta = \Delta_1$

$N = N_1 + N_\Delta + N_3)$

$$f_{\Lambda_2}(x_1,N_1,x_\Delta,N_\Delta,x_3,N_3|x_{\Lambda_2})\frac{dx_1\ dx_\Delta\ dx_3}{N!}$$

$$= \frac{e^{-\beta(H_1(x_1) + H_\Delta(x_\Delta) + H_3(x_3)) + \beta\mu(N_1 + N_\Delta + N_3)}}{G_{\Lambda_2}}\ \frac{dx_1\ dx_\Delta\ dx_3}{N!}$$

(H_1 and H_3 include the interaction energy with x_{Δ_1} and x_{Δ_2} respectively.)
The conditional density of x_1, N_1 given x_{Δ_1} and x_{Δ_2} is hence

$$\frac{\displaystyle\sum_{N_3}\left(\frac{N!}{N_1!\ N_\Delta!\ N_3!}\right)\int f_{\Lambda_2}(\ldots)\frac{dx_3}{N!}\ dx_1}{\displaystyle\sum_{N_1 N_3}\left(\frac{N!}{N_1!\ N_\Delta!\ N_3!}\right)\int f_{\Lambda_2}(\ldots)\frac{dx_1\ dx_3}{N!}}$$

In this expression the factors

$$G_{\Lambda_2}^{-1}\ \sum_{N_3}\ \int e^{-\beta(H_\Delta(x_\Delta) + H_3(x_3)) + \beta\mu(N_\Delta + N_3)}\ \frac{dx_3}{N_\Delta!\ N_3!}$$

are cancelled in numerator and denominator, so the conditional density
is

$$\frac{e^{-\beta H_1(x_1) + \beta\mu N_1}}{G_{\Lambda_1}(\beta,\mu|x_\Delta)}\ \frac{dx_1}{N_1!}\ ,$$

i.e. equal to the g.can. one with the parameters $\beta = \beta_2$, $\mu = \mu_2$ belonging
to the larger region. This argument shows that any two regions have the
same β and μ since the values have to be equal to β and μ of a sufficiently large Λ containing both of them.

Let us now collect the formulas for the g.can. density of a system in Λ
given the configuration y outside Λ:

Lemma 3. If the probability density for x,N in Λ in a large surrounding
system has a limit in the thermodynamic limit of the surrounding system,
then it has to be a g.can. density:

$$f(x,N|y)\frac{dx}{N!} = \frac{e^{-\beta H(x|y) + \beta\mu N}}{G_\Lambda(\beta,\mu|y)}\ \frac{dx}{N!}\quad\text{with}$$

$$G_\Lambda(\beta,\mu|y) = \sum_N \int e^{-\beta H(x|y) + \beta\mu N} \frac{dx}{N!}$$

$$= \sum_N \int e^{-\beta E + \beta\mu N} \Omega'_\Lambda(E,N|y) \; dE = \sum_N e^{\beta\mu N} Z_\Lambda(\beta,N|y).$$

The density of E,N is hence given by

$$\frac{e^{-\beta E + \beta\mu N} \Omega'_\Lambda(E,N|y) \; dE}{G_\Lambda(\beta,\mu|y)} \quad , \text{ and that of N by } \quad \frac{e^{\beta\mu N} Z_\Lambda(\beta,N|y)}{G_\Lambda(\beta,\mu|y)} \quad .$$

The average values of E,N are given by

$$\langle E \rangle = - \frac{\partial \log G_\Lambda(\beta,\mu|y)}{\partial\beta} \qquad \text{(Derivation with } \beta\mu = \text{const.)}$$

$$\langle N \rangle = \frac{1}{\beta} \frac{\partial \log G_\Lambda(\beta,\mu|y)}{\partial\mu}$$

The conditional density of x given E,N is the m.can. one, and given only N is the can. one defined by β,N.

We have in the derivation assumed that the interaction is of finite range, but the above formulas can make sense also for interactions of infinite range.

Some applications of the grand canonical distribution

1. The equation of state of an arbitrary gas in a slowly varying
    external field

Let us consider a gas influenced also by an external field of the form
$U(\frac{q}{L})$ giving a contribution $\sum_i U(\frac{q_i}{L})$ to the total energy. Here $U(x)$
is a nice function of $x \in R^3$, and we shall consider the situation when
$L \to \infty$, so that U varies very slowly on a microscopic scale. We shall
however consider the density, pressure etc. on a scale $q = L \cdot x$, on
which there is a non trivial variation of these macroscopic quantities.
Consider now the state in a macroscopically infinitesimal cell Λ with
side $L \cdot \Delta x$ centered at $L \cdot x$. The potential has the essentially constant
value $U(x)$ in the cell, so its contribution to the energy in the cell
is $N \cdot U(x)$. The partition function G for the cell is hence
$G_\Lambda(\beta,\mu-U(x)|y)$.

Now, as we shall prove later, it is a basic fact that also $g_\Lambda(\beta,\mu|y) =$
$= \frac{1}{|\Lambda|} \log G_\Lambda(\beta,\mu|y)$ is extensive as $\Lambda \to \infty$, so that
$g(\beta,\mu) = \lim_\Lambda g_\Lambda(\beta,\mu|y)$ exists, and is independent of the boundary
effect coming from y.

Moreover, since g_Λ are convex in β,μ it follows that the derivatives
of g_Λ converge to those of g , when these exist (which happens
except at most for denumerably many values of β,μ.). Hence we can
conclude that the average density at x

$$n_\Lambda(x) = \beta^{-1} \frac{\partial g_\Lambda(\beta,\mu-U(x)|y)}{\partial \mu} \quad \text{converges to}$$

$$n(x) = \beta^{-1} \frac{\partial g(\beta,\mu-U(x))}{\partial \mu} \quad \text{as } L \to \infty \text{ with } \Delta x \text{ fixed.}$$

This is the general barometric formula for a non ideal gas. To see how
the pressure can be expressed in terms of $g(\beta,\mu)$ let us independently
of our previous formulas define it by the macroscopic balance equation

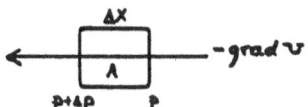

The increase in p in the direction of the force $-$ grad U balances it:

$$(\Delta x)^2 (p(x) - p(x + \Delta x)) = n(x) (\Delta x)^3 \cdot \text{grad } U(x).$$

Hence $p(x)$ is defined by the equation

$$\text{grad } p(x) = - n(x) \text{ grad } U(x) = - \beta^{-1} g_\mu'(\beta,\mu-U(x)) \text{ grad } U(x) =$$

$$= \beta^{-1} \text{ grad}_x \ g(\beta,\mu-U(x)), \text{ which when integrated gives}$$

$$p(x) = \beta^{-1} g(\beta,\mu-U(x)) \text{ if } p = 0 \text{ when } g = 0.$$

This is the case if we have a situation when $U(x) \to \infty$ sufficiently rapidly when $|x| \to \infty$ so that g and $n \to 0$ when $|x| \to \infty$ and $p(\infty) = 0$, i.e. the gas in enclosed in a "potential well".

To see the relation between this and our previous definition of p,

$$p = \beta^{-1} \frac{\partial \log Z_\Lambda(\beta,N)}{\partial V} \quad \text{remember that} \quad G_\Lambda(\beta,\mu) = \sum_N e^{\beta\mu N} Z_\Lambda(\beta,N)$$

so that at least formally

$$\frac{\partial G_\Lambda(\beta,\mu)}{\partial V} = \sum_N e^{\beta\mu N} \frac{\partial Z_\Lambda(\beta,N)}{\partial V} =$$

$$= \sum_N e^{\beta\mu N} Z_\Lambda(\beta,N) \frac{\partial \log Z_\Lambda(\beta,N)}{\partial V} = G_\Lambda(\beta,\mu)\beta \ \langle p(\beta,N)\rangle$$

$$\therefore \quad \frac{\partial \log G_\Lambda(\beta,\mu)}{\partial V} = \beta \ \langle p(\beta,N)\rangle \approx \beta p(\beta, \langle N\rangle)$$

if N has a sharp distribution when $\Lambda \to \infty$. We have $\log G_\Lambda(\beta,\mu) = V \cdot g(\beta,\mu) + o(V)$ as $\Lambda \to \infty$, so if this relation can be differentiated we have $g = \beta p$ in the limit $\Lambda \to \infty$. (It is not very simple to complete this argument to a rigorous proof.)

The equations

$$\beta p = g(\beta,\mu-U(x))$$
$$\beta n = g_\mu'(\beta,\mu-U(x))$$

define the equation of state of the system in parametric form, p,n as functions of β,μ. If μ is eliminated the relation between p,n,β is obtained.

The parametric form is however very convenient to use. Consider e.g. the situation when $U(x) = mgx_1$, a gas in a constant gravitational field:

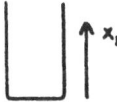

g is a convex increasing function of μ, and is hence differentiable except when the derivative possibly jumps.

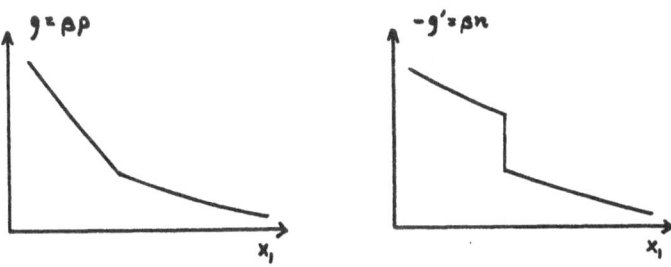

In an interval where g_μ' exists p and n are decreasing continuous functions of x_1, and if $g_\mu'(\beta,\mu)$ makes a jump at a critical value μ_c then the density suddenly drops when $\mu - gx_1 = \mu_c$ but the pressure is continuous at this point. This indicates that a phase transition takes place, the system consists of a heavy phase (liquid) and above it a light phase (gas) in equilibrium with each other. They are separated by the gravitational field and have a sharp interphase at $\mu - gx_1 = \mu_c$. An important and interesting problem is to investigate when such phase transitions can take place.

2.   The equilibrium rule for a binary chemical reaction

Let us consider such a reaction $A + B \rightleftarrows AB$ where the molecules A,B,AB do not interact except when they are formed. In the reaction $\rightarrow$ one A- and one B-molecule form an AB-molecule and an amount of energy U is required. Suppose that the state of a single molecule of the respective types can be described by some coordinates x,y,z resp. (e.g. describing vibrations, rotations, positions etc.). The state of a gas consisting of N_1, N_2, N_3 particles respectively is then described by fixing

$(x_1, \ldots x_{N_1}, y_1, \ldots y_{N_2}, z_1, \ldots z_{N_3})$ modulo permutations of identical par-

ticles. The energy of such a state is $\sum\limits_1^{N_1} H_1(x_i) + \sum\limits_1^{N_2} H_2(y_i) + \sum\limits_1^{N_3} H_3(z_i) +$

$+ U \cdot N_3 = H(x,y.z) + U \cdot N_3$.

$H_1(x_1)$ etc. is the energy connected with the inner degrees of freedom
of the molecules, and there is no interaction among these. The can. partition
function for all states with N_1, N_2, N_3 given is hence:

$$Z(\beta, N_1, N_2, N_3) = \int e^{-\beta\sum\limits_1^{N_1} H_1(x_i)}\, e^{-\beta\sum\limits_1^{N_2} H_2(y_i)}\, e^{-\beta\sum\limits_1^{N_3} H_3(z_i) - \beta U N_3}\, \frac{dx}{N_1!}\, \frac{dy}{N_2!}\, \frac{dz}{N_3!} =$$

$$\frac{Z_1^{N_1}}{N_1!}\, \frac{Z_2^{N_2}}{N_2!}\, \frac{Z_3^{N_3}}{N_3!}\, e^{-\beta U N_3} \quad , \quad \text{where} \quad Z_1 \quad \text{etc. are the partition functions}$$

of a single molecule

$$Z_1 = \int e^{-\beta H_1(x)}\, dx \quad \text{etc.}$$

The can. partition function for a gas with M_1 and M_2 A- and B-molecules
respectively is hence

$$Z(M_1, M_2) = \sum_{N=0}^{M_1 \wedge M_2} Z(M_1 - N, M_2 - N, N)$$

(To form 1 AB 1 A and 1 B is needed.)

The probability density for N_3 is hence

$$p(N_3 = N) = \frac{Z(M_1 - N, M_2 - N, N)}{Z(M_1, M_2)} \quad , \quad \text{and is not so easy to handle.}$$

If we now consider a small subvolume of the system and ask for the equilibrium concentrations of A,B,AB, we can use the g. can. distribution
where $M_1 M_2$ are also stochastic. In this case we must use two potentials
μ_1, μ_2 and get:

$$G(\beta,\mu_1,\mu_2) = \sum_{M_1 M_2} e^{\beta\mu_1 M_1 + \beta\mu_2 M_2} Z(M_1,M_2) =$$

$$= \sum_{N_1,N_2,N_3} e^{\beta\mu_1(N_1+N_3)+\beta\mu_2(N_2+N_3)} \frac{Z_1^{N_1}}{N_1!} \frac{Z_2^{N_2}}{N_2!} \frac{Z_3^{N_3}}{N_3!} e^{-\beta U N_3},$$

so we get the simple formula

$$\log G = Z_1 e^{\beta\mu_1} + Z_2 e^{\beta\mu_2} + Z_3 e^{\beta(\mu_1+\mu_2-U)}$$

μ_1 and μ_2 are determined by the equations

$$\begin{cases} M_1 = \beta^{-1} \dfrac{\partial \log G}{\partial \mu_1} = Z_1 e^{\beta\mu_1} + Z_3 e^{\beta(\mu_1+\mu_2-U)} \\[3mm] M_2 = \beta^{-1} \dfrac{\partial \log G}{\partial \mu_2} = Z_2 e^{\beta\mu_1} + Z_3 e^{\beta(\mu_1+\mu_2-U)} \end{cases}$$

From the formula for G we then also get:

$$\begin{cases} \langle N_1 \rangle = Z_1 e^{\beta\mu_1} \\[2mm] \langle N_2 \rangle = Z_2 e^{\beta\mu_2} \\[2mm] \langle N_3 \rangle = Z_3 e^{\beta(\mu_1+\mu_2-U)}, \end{cases}$$

so the following relation is valid

$$\frac{\langle N_1 \rangle \langle N_2 \rangle}{\langle N_3 \rangle} = \frac{Z_1(\beta) Z_2(\beta)}{Z_3(\beta)} e^{\beta U} \equiv k(\beta)$$

independently of μ_1,μ_2. This is the famous law of mass action, which shows that the equilibrium concentrations are determined by the above relation and

$$\begin{cases} M_1 = \langle N_1 \rangle + \langle N_3 \rangle \\[2mm] M_2 = \langle N_2 \rangle + \langle N_3 \rangle \end{cases}$$

if M_1,M_2 are given. We also see that the equilibrium constant $k(\beta)$

can be computed from the knowledge of the inner structure of the molecules, and that it has a simple dependence on U, the dissociation energy. In this case it is easy to verify directly that the variances of N_i are proportional to $\langle N_i \rangle$, so the law of large numbers holds with great precision.

3. The pressure in a mixture of independent particles, partial pressure and osmotic pressure

If as in the previous example we have a system of two types of particles A and B we see like in 1. that locally we have for the pressure

$$\text{grad } p(x) = -n_1(x) \text{ grad } U_1(x) - n_2(x) \text{ grad } U_2(x)$$

$$\therefore \quad \beta \text{ grad } p = -\frac{\partial g_1(\beta,\mu_1-U_1)}{\partial \mu_1} \text{ grad } U_1(x) - \frac{\partial g_2(\beta,\mu_2-U_2)}{\partial \mu_2} \text{ grad } U_2(x) =$$

$$= \text{grad } (g_1+g_2) , \quad \text{and hence}$$

$$\beta p(x) = g_1(\beta,\mu-U_1(x)) + g_2(\beta,\mu-U_2(x)) = \beta p_1(x) + \beta p_2(x).$$

p is hence the sum of the "partial pressures" of the components.

An example of this is the occurrence of "osmotic pressure". Consider a system consisting of a vessel with two parts separated by a semi-permeable wall preventing B- but not A-molecules from passing it:

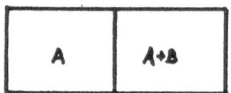

The pressure to the left is $p_1 = \beta^{-1} g_1$ and to the right $p_r = \beta^{-1}(g_1+g_2)$. (The A-particles do not feel the wall, so g_1 is computed by integrating over the whole volume, and μ_1 is constant there, whereas g_2 is obtained by integrating over the right half.) There is hence a pressure difference $\Delta p = \beta^{-1} g_2$ at the wall. It is called the osmotic pressure. If the B-particles form an ideal gas it is hence determined by $\Delta p = n_B \cdot kT$, the same formula as for an ideal gas.

3. The law of large numbers for macroscopic variables and the foundations
   of thermodynamics

3.1. General study of the probability laws of macrovariables

We are now going to give a general argument showing that macroscopic
variables like the total energy, number of particles etc. have very
sharp distributions around their average values in the thermodynamic
limit. In this process we will also see to what extent the various
"ensembles" introduced give the same results. We are also going to see
that the rules for how these averages are related and vary when external
conditions are changed are those prescribed by the laws of thermodynamics.

Let us again state our basic microscopic description of a system in a
container $\Lambda \subset R^d$. To begin with we will consider only the part of the
system depending on the position variables q_i and afterwards add the
part depending on the p_i for simplicity. We also restrict our atten-
tion to systems of one type of identical particles. For any value N of
the no. of particles in Λ we have the state space Λ^N of possible
states $q = (q_1, \ldots, q_N) \subset R^{dN}$ with all $q_i \in \Lambda$, and the basic measure
$$\omega_N(dq) = \frac{dq_1 \ldots dq_N}{N!} .$$ (If the particles are rigid spheres of radius
$r/2$ we should instead take $\omega_N(dq)$ as the restriction of the above measure
to the region $\subset \Lambda^N$ where $|q_i - q_j| > r$ for all $i \neq j$.) The state
space for the system with a variable no. of particles is then $\Gamma_\Lambda = \bigcup_0^\infty \Lambda^N$,
with $\omega(dq)$ defined on Γ_Λ by saying that its restriction to Λ^N
is $\omega_N(dq)$ for all N. (For $N = 0$ we let Λ^0 consist of a single point
λ and take $\omega\{\lambda\}) = 1.$)

A point of Γ_Λ is hence $q = (q_1, \ldots, q_N)$ with N arbitrary ≥ 0. A typical macroscopic variable (observable) is e.g. the total potential energy
$$v(q) = \sum_{1 \leq i < j \leq N} v(q_i - q_j) \text{ if } q \in \Lambda^N ,$$ the total no. of particles in a
region $\Lambda' \subset \Lambda$, $\sum_1^N \chi_{\Lambda'}(q_i)$ if $q \in \Lambda^N$ etc.

More generally, for any symmetric function of m variables $u(q_1, \ldots, q_m)$
we can define $U(q) = \sum_{(i_1, \ldots, i_m) \subset \{1, \ldots, N\}} u(q_{i_1}, \ldots, q_{i_m})$ for $q \in R^{dN}$.

This will be a symmetric function of $(q_1, \ldots, q_N)$, and if u is transla-
tionally invariant, so that it depends only on $(q_2 - q_1, \ldots, q_m - q_1)$, then it

will also be invariant when the q_i are all shifted by the same amount to $q_i + q$ $q \in R^d$. If moreover $u(q_1, \ldots, q_m)$ is of finite range R, so that $u(q_1, \ldots, q_m) = 0$ as soon as $|q_i - q_j| > R$ for some $i \neq j$ then $U(q)$ will have the property that if q can be split into (q', q'') with $|q_i' - q_j''| > R$ for all i, j then $U(q) = U(q') + U(q'')$.

We will consider observables $U(q)$ defined for finite configurations q having these properties:

a) U is symmetric when particles are permuted.

b) U is invariant under translations of the configurations in R^d.

c) U is of finite range $R \geq r$, as defined above.

This last assumption can be relaxed, but it simplifies the arguments considerably.

We will consider a situation where we are interested in a finite no. of such observables $U_1, \ldots, U_M$ which describe our system macroscopically. A general m. can. ensemble can be defined by considering all $q \in \Gamma_\Lambda$ having fixed values of $U_i(q)$ as equally probable (according to $\omega(dq)$), and a general canonical ensemble by saying that q has a probability law proportional to $e^{-\sum_1^M a_i U_i(q)} \omega(dq)$, where $(a_1, \ldots, a_M)$ are some parameters. (Inverse temperature etc.) (We use vector notation $U(q) = (U_1(q), \ldots, U_M(q))$ $a \cdot U = \sum_1^M q_i U_i$ etc.)

These probability laws are expressed in terms of the "structure measure":

$$\Omega_\Lambda(A) = \omega(\tfrac{U(q)}{|\Lambda|} \in A , q \in \Gamma_\Lambda) \quad \text{for} \quad A \subset R^M ,$$

$$\Omega_\Lambda(A) = \sum_{N=0}^{\infty} \int_{\substack{\frac{U(q)}{|\Lambda|} \in A \\ q \in \Lambda^N}} \omega_N(dq) .$$

We also introduce the corresponding canonical measure:

$$\Omega_\Lambda(A, a) = \int_{\substack{\frac{U(q)}{|\Lambda|} \in A \\ q \in \Gamma_\Lambda}} e^{-a \cdot U(q)} \omega(dq) = \int_A e^{-|\Lambda|(a \cdot u)} \Omega_\Lambda(du)$$

$$(\Omega_\Lambda(A, o) = \Omega_\Lambda(A))$$

We will use a slightly different concept of m. can. measure than before and say that $\frac{U(q)}{|\Lambda|}$ is not exactly fixed to a value $u \in R^M$, but require that $\frac{U(q)}{|\Lambda|} \in \Delta$, where Δ is a small neighbourhood of $u \in R^M$ (a thick "energy shell"). We are then going to consider the limit of the m. can. distribution when first $\Lambda \to \infty$ and then $\Delta \to u$. (This is technically much simpler than considering the thin energy shell and physically very reasonable, since it is difficult to keep a macroscopic variable constant with microscopic precision.)

The m. can. probability distribution of an observable $\frac{U_o(q)}{|\Lambda|}$ defined by the restriction $\frac{U(q)}{|\Lambda|} \in \Delta$ can then be expressed in terms of Ω_Λ as follows:

Include U_o among $U_1, \ldots, U_M$ and define $\Omega_\Lambda(A \times \Delta) = \omega(\frac{U_o(q)}{|\Lambda|} \in A, \frac{U(q)}{|\Lambda|} \in \Delta)$ for $A \subset R^1$. The probability distribution of $\frac{U_o}{|\Lambda|}$ is then:

$$P_\Lambda(\frac{U_o(q)}{|\Lambda|} \in A | \Delta) = \frac{\Omega_\Lambda(A \times \Delta)}{\Omega_\Lambda(R^1 \times \Delta)} = \frac{\Omega_\Lambda(A \times \Delta)}{\Omega_\Lambda(\Delta)} \ .$$

(We use the same letter Ω_Λ for both measures and let the difference be clear from the argument $A \times \Delta, \Delta$ etc.)

The can. probability distribution of $\frac{U(q)}{|\Lambda|}$ is expressed by:

$$P_\Lambda(\frac{U(q)}{|\Lambda|} \in A | a) = \frac{\Omega_\Lambda(A, a)}{\Omega_\Lambda(R^M, a)} \qquad A \subset R^M$$

If we want to consider another observable $\frac{U_o(q)}{|\Lambda|}$ as well as above we get:

$$P_\Lambda(\frac{U_o(q)}{|\Lambda|} \in A, \frac{U(q)}{|\Lambda|} \in B | a) = \frac{\Omega_\Lambda(A \times B, (o, a))}{\Omega_\Lambda(R^{M+1}, (o, a))} =$$

$$= \frac{\Omega_\Lambda(A \times B, (o, a))}{\Omega_\Lambda(R^M, a)} \ , \qquad \begin{matrix} A \subset R^1 \\ B \subset R^M \end{matrix} \qquad \text{with the same abuse of notation.}$$

($U(q)$ will be an extensive quantity, and therefore we consider all the time $\frac{U(q)}{|\Lambda|}$ etc. whose distribution will obey the law of large numbers as $\Lambda \to \infty$.) As before we put

$$G_\Lambda(a) = \exp|\Lambda| g_\Lambda(a) = \Omega_\Lambda(R^M, a)$$

The most important special case is when $U_1(q)$ = the total potential energy, $U_2(q) = N(q)$ = the total no. of particles. Then $a_1 = \beta$, $a_2 = -\beta\mu$, and the analog of $\Omega_\Lambda(E,N)$ defined earlier is $\Omega_\Lambda(A)$ with A defined by $u_1 \leq \frac{E}{|\Lambda|}$ $u_2 = \frac{N}{|\Lambda|}$.

Now, the fundamental asymptotic properties of the probabilities is a consequence of the fact that $\log \Omega_\Lambda(A,a)$ is extensive:

Theorem 2.   As $\Lambda \to \infty$

$$\lim_{\Lambda \to \infty} \frac{1}{|\Lambda|} \log \Omega_\Lambda(A,a) = s(A,a) \quad \text{exists, if } A \text{ is any open convex subset of } R^M .$$

The values $\pm\infty$ are not excluded.

For $a = o$ we write

$$s(A) = \lim_{\Lambda \to \infty} \frac{1}{|\Lambda|} \log \Omega_\Lambda(A)$$

$s(A,a) = -\infty$ iff $s(A) = -\infty$, which happens iff $\Omega_\Lambda(A) = o$ for all Λ. $s(A,a)$ will have the following regularity property:

$$.s(A,a) = \sup_{C \subset A} s(C,a) , \quad \text{where } C \text{ is open convex with compact closure } \subset A.$$

We will give the proof of Theorem 2 later when we have seen its consequences. The statement $\Lambda \to \infty$ has to be made precise. At this stage let us assume that Λ are rectangular boxes all of whose sides become infinite.

We shall now study the set function $s(A,a)$ and see how it depends on A .

Lemma 4. $s(A,a)$ has the following properties: $(A = \text{open convex} \subset R^M)$

a) $s(A) \leq 1$, and $s(A,a) \leq 1 + \sup_{u \in A}(-a \cdot u)$,
   so $s(A,a)$ is bounded on any bounded A .

b) $s(A,a) \leq s(B,a)$ if $A \subseteq B$.

c)   If $A = \overset{n}{\underset{1}{U}} A_i$, and A_i (but not necessarily A) are open convex, then $s(A,a)$ is also defined, and $s(A,a) = \max_i s(A_i,a) \equiv \overset{n}{\underset{1}{V}} s(A_i,a)$

d) $s(\frac{A_1 + A_2}{2}, a) \geq \frac{s(A_1,a) + s(A_2,a)}{2}$

   $(\frac{A_1 + A_2}{2} = \{u = \frac{u_1 + u_2}{2} \text{ with } u_1 \in A_1, u_2 \in A_2\})$.

Proof:

a) $\Omega_\Lambda(A) \leq \sum_o^\infty \int_{q\in\Lambda} N \frac{dq}{N!} = \sum_o^\infty \frac{|\Lambda|^N}{N!} = e^{|\Lambda|}$,

so $s(A) \leq 1$. $\Omega_\Lambda(A,a) \leq \Omega_\Lambda(A) \exp|\Lambda| \sup_{u\in A}(-a\cdot u)$, which gives the
bound for $s(A,a)$.

b) is clear.

c) If $A = A_1 \cup A_2$, with $s(A_1,a) \geq s(A_2,a)$ e.g. we have for any $\varepsilon > o$
$\Omega_\Lambda(A_1,a) \leq \Omega_\Lambda(A,a) \leq \Omega_\Lambda(A_1,a) + \Omega_\Lambda(A_2,a) \leq 2 \exp|\Lambda|(s(A_1,a) + \varepsilon)$
if Λ is big enough.

$\therefore$ $s(A_1,a) \leq s(A,a) \leq s(A_1,a) + \varepsilon$ for any $\varepsilon > o$, and $s(A,a) = s(A_1,a)$.

For $A = \overset{n}{\underset{1}{\cup}} A_i$ the proof goes by induction.

d) For any Λ let Λ' consist of two translates of Λ and a corridor, C,
of width R separating them:

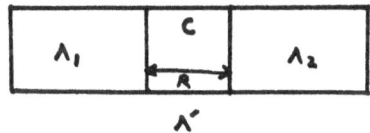

For any pair of configurations q_1, q_2 in Λ_1 and Λ_2 with
$\frac{U(q_1)}{|\Lambda|} \in B_1$, $\frac{U(q_2)}{|\Lambda|} \in B_2$, and no. of particles N_1 and N_2 the combined
configuration q is allowed in Λ' and has

$\frac{U(q)}{|\Lambda'|} = \frac{U(q_1) + U(q_2)}{|\Lambda'|} \in \frac{|\Lambda|}{|\Lambda'|} (B_1 + B_2) \equiv B'$ and $N = N_1 + N_2$

Hence $\underset{\substack{\frac{U(q)}{|\Lambda'|}\in B' \\ q\in\Lambda'^N}}{\int} e^{-a\cdot U(q)} dq \geq \underset{\substack{\frac{U(q)}{|\Lambda'|}\in B' \\ q\in(\Lambda'\smallsetminus C)^N}}{\int} e^{-a\cdot U(q)} dq$

$\geq \underset{N_1+N_2=N}{\sum} \binom{N}{N_1} \underset{\substack{\frac{U(q_1)}{|\Lambda_1|}\in B_1 \\ q_1\in\Lambda_1^{N_1}}}{\int} e^{-a\cdot U(q_1)} dq_1 \underset{\substack{\frac{U(q_2)}{|\Lambda_2|}\in B_2 \\ q_2\in\Lambda_2^{N_2}}}{\int} e^{-a\cdot U(q_2)} dq_2$

(There are $\binom{N}{N_1}$ ways of chosing N_1 particles to Λ_1 .)

Remembering that $\omega_N(dq) = \frac{dq}{N!}$, we get

$$\sum_{\substack{N \\ \frac{U(q)}{|\Lambda'|} \in B' \\ q \in \Lambda, N}} \int e^{-a \cdot U(q)} \omega_N(dq) \geq$$

$$\geq \sum_{N_1, N_2} \int_{\substack{\frac{U(q_1)}{|\Lambda_1|} \in B_1 \\ q_1 \in \Lambda_1^{N_1}}} e^{-a \cdot U(q_1)} \omega_{N_1}(dq_1) \int_{\substack{\frac{U(q_2)}{|\Lambda_2|} \in B_2 \\ q_2 \in \Lambda_2^{N_2}}} e^{-a \cdot U(q_2)} \omega_{N_2}(dq_2)$$

I.e.

$$\Omega_{\Lambda'}(B ,a) \geq \Omega_\Lambda(B_1,a) \, \Omega_\Lambda(B_2,a)$$

Now, as $\Lambda \to \infty$ $\frac{|\Lambda|}{|\Lambda'|} \uparrow \frac{1}{2}$, and B' is nearly $\frac{B_1 + B_2}{2}$, but an extra approximation argument is needed:

Theorem 2 tells us that for any $\varepsilon > 0$ we can find $B_i \subset A_i$ with $\bar{B}_i$ compact $\subset A_i$ and $s(B_i,a) > s(A_i,a) - \varepsilon$. Then $\bar{B} = \frac{\bar{B}_1 + \bar{B}_2}{2} \subset \frac{A_1 + A_2}{2}$ is also bounded, and for any point $u' \in B'$ we have $u' = \frac{2|\Lambda|}{|\Lambda'|} u$ with $u \in B$, so $d(u',B) \leq (1 - \frac{2|\Lambda|}{|\Lambda'|}) \max_{u \in B} |u| \to 0$ uniformly in u'. Hence $d(B',\bar{B}) \to 0$, and $B' \subset \frac{A_1 + A_2}{2}$ if Λ is big enough, because $\bar{B}$ being compact has a strictly positive distance to the complement of $\frac{A_1 + A_2}{2}$ which is closed:

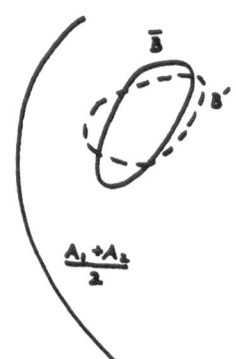

We see that

$$\Omega_{\Lambda'}(\frac{A_1 + A_2}{2}, a) \geq \Omega_{\Lambda}(B_1,a)\, \Omega_{\Lambda}(B_2,a)$$

when Λ is big, and hence

$$2s(\frac{A_1 + A_2}{2}, a) \geq s(B_1,a) + s(B_2,a) \geq s(A_1,a) + s(A_2,a) - 2\varepsilon$$

for all $\varepsilon > o$, which proves d).

Property c) suggests that if A is partitioned into small cells A_i centered at u_i then $s(A,a) = \overset{n}{\underset{1}{V}} s(A_i,a) \approx \overset{n}{\underset{1}{V}} s(u_i,a)$, if $s(A_i,a)$ is approximatively equal to a constant value $s(u_i,a)$ for a small cell $A_i \ni u_i$. This is indeed true:

Lemma 5. $s(A,a)$ can be expressed as $s(A,a) = \sup\limits_{u \in A} s(u,a)$, for $A =$ open convex, with $s(u,a)$ defined by

$s(u,a) = \inf\limits_{A \ni u} s(A,a)$. (This definition makes sense also if some components of u take values $\pm\infty$, i.e. when $u \in \bar{R}^M$, where $\bar{R}$ is the extended real axis $\{-\infty\} \cup R \cup \{+\infty\}$.) $s(u,a)$ has the following properties:

a) $s(u,a) = s(u,0) - a \cdot u \equiv s(u) - a \cdot u$

b) $s(u)$ is ≤ 1, upper semicontinuous (u.s.c.) (also on the extended domain $\bar{R}^M$) and concave (possibly $= -\infty$ for some u).

c) $s(u,a)$ is uniquely determined in the following sense: If $s(A,a) = \sup\limits_{u \in A} \tilde{s}(u,a)$ for all open convex A, then $\tilde{s}(u,a) = s(u,a)$ if $\tilde{s}$ is u.s.c. at u.

d) $D = \{u; u \in R^M, s(u) > -\infty\}$ is convex, and its closure $\bar{D}$ is given by

$$\bar{R} = \underset{\Lambda}{U} \text{ (essential range of } \frac{U(q)}{|\Lambda|} \text{ when } q \in \Gamma_{\Lambda})$$

Proof: Since $s(A,a)$ is decreasing in A $s(u,a) = \lim\limits_{A_i \to u} s(A,a)$ for any sequence of open convex A_i shrinking to u. (An open neighbourhood of $+\infty$ in $\bar{R}$ e.g. is an interval $A = (a,\infty)$ etc.) Clearly $s(A,a) \geq \sup\limits_{u \in A} s(u,a)$ from the definition of $s(u,a)$, so if $s(A,a) = -\infty$ there is equality. If $s(A,a) > -\infty$ take $\varepsilon > o$ arbitrary and $\varepsilon' < s(A,a) - \varepsilon$. For any $u \in A$ with $s(u,a)$ finite there is an open $A_u \ni u$ such that $s(A_u,a) < s(u,a) + \varepsilon$, and if $s(u,a) = -\infty$ there is an $A_u \ni u$ such that $s(A_u,a) < \varepsilon'$. Hence these A_u form an open covering of A. Take a C with

$\bar{C}$ compact $\subset A$ such that $s(C,a) > s(A,a) - \varepsilon$. Since $\bar{C}$ is compact it can be covered by a finite subcovering $\{A_{u_i}\}_1^n$. Hence

$$\varepsilon' < s(C,a) \leq s(\bigcup_1^n A_{u_i},a) = \bigvee_1^N s(A_{u_i},a) \,,$$

and we see that $\bigvee_1^n$ has to be attained for some u_i with $s(u_i,a) > -\infty$ and hence

$$s(A_{u_i},a) < s(u_i,a) + \varepsilon \leq \sup_{u\in A} s(u,a) + \varepsilon, \quad \text{so}$$

$$s(A,a) < s(C,a) + \varepsilon < \sup_{u\in A} s(u,a) + 2\varepsilon \quad \text{for any} \quad \varepsilon > o \quad \text{and}$$

$$s(A,a) = \sup_{u\in A} s(u,a) \,.$$

a) If A is a neighbourhood of u of diameter $\leq \varepsilon$ then since

$$\Omega_\Lambda(A,a) = \int_A e^{-|\Lambda|(a\cdot u)} \Omega_\Lambda(du) \quad \text{we have}$$

$$\Omega_\Lambda(A) \, e^{-|\Lambda|((a\cdot u)+\varepsilon|a|)} \leq \Omega_\Lambda(A,a) \leq \Omega_\Lambda(A) \, e^{-|\Lambda|((a\cdot u)-\varepsilon|a|)}$$

and hence

$$s(A) - (a\cdot u) - \varepsilon|a| \leq s(A,a) \leq s(A) - (a\cdot u) + \varepsilon|a|$$

and a) is proved by letting A shrink to u .

b) $s(u) \leq 1$ follows from $s(A) \leq 1$.

Upper semicontinuity means that the "Epigraph" = $\{(s,u) \in \bar{\mathbb{R}}^{M+1} \,; \, s \leq s(u)\}$ is closed, i.e. its complement is open i.e. if $s > s(u)$ then there is an open $A \ni u$ and $\varepsilon > o$ such that $(s-\varepsilon,s+\varepsilon)\times A \ni (s,u)$ is also in the complement, or $s(u') \leq s-\varepsilon$ if $u' \in A$. This property is clear, because if $s > s(u) = \inf_{A\ni u} s(A)$ then there is an $A \ni u$ and $\varepsilon > o$ such that $s(A) \leq s-\varepsilon$, so that $s(u') \leq s(A) \leq s-\varepsilon$ for $u' \in A$.

$s(\dfrac{u_1 + u_2}{2}) \geq \dfrac{s(u_1) + s(u_2)}{2}$ follows from Lemma 4 d) by letting A_1,A_2 shrink to u_1,u_2. It then follows that $s(\lambda u_1 + (1 - \lambda)u_2) \geq$ $\geq s(u_1) + (1 - \lambda) s(u_2)$ if $0 \leq \lambda \leq 1$ and $\lambda = a$ dyadic rational, and for λ arbitrary $\in [0,1]$ it follows by the semicontinuity.

c) If $s(A,a) = \sup_{u\in A} \tilde{s}(u,a)$ for all A then $s(A,a) \geq \tilde{s}(u,a)$ if $A \ni u$,

so $s(u,a) \geq \tilde{s}(u,a)$. If $s(u,a)$ were $> \tilde{s}(u,a)$, and $\tilde{s}$ u.s.c. at u

then there would be a neighbourhood $A \ni u$ and $\varepsilon > o$ such that $\tilde{s}(u',a) \leq s(u,a) - \varepsilon$ in A , but then we would have $s(A,a) = \sup_{u \in A} \tilde{s}(u,a) < s(u,a)$, contradicting $s(u,a) = \inf_{A \ni u} s(A,a)$.

d) The proof is given after the proof of Theorem 2 on p. 93.

We shall now see how the function $s(u)$ can be used to study the asymptotic form of the m. can. and can. distributions for typical macroscopic variables $\frac{U(q)}{|\Lambda|}$. Consider first the m. can. distribution defined on p. 41. We have

$$\frac{1}{|\Lambda|} \log P_{\Lambda}(\frac{U_o(q)}{|\Lambda|} \in A | \Delta) = \frac{1}{|\Lambda|} \log \Omega_{\Lambda}(A \times \Delta) - \frac{1}{|\Lambda|} \log \Omega_{\Lambda}(R^1 \times \Delta)$$

$$\rightarrow s(A \times \Delta) - s(R^1 \times \Delta) = \sup_{\substack{u_o \in A \\ u \in \Delta}} s(u_o,u) - \sup_{\substack{u_o \\ u \in \Delta}} s(u_o,u)$$

If we now let Δ shrink to u we filter out the value of $s(u_o,u)$:

<u>Lemma 6.</u> If $s(u) > -\infty$ then

$$\lim_{\Delta \to u} \lim_{\Delta \to \infty} \frac{1}{|\Lambda|} \log P_{\Lambda}(\frac{U_o(q)}{|\Lambda|} \in A | \Delta) =$$

$$= \sup_{u_o \in A} s(u_o,u) - \sup_{u_o} s(u_o,u) \equiv s(A|u) - s(R|u)$$

if these are u.s.c. at u , and then $s(R|u) = s(u)$. This happens if $s(A|u) = s(\bar{A}|u)$, $s(R|u) = s(\bar{R}|u)$, with $\bar{A} = [a',a'']$ if $A = (a',a'')$ etc. For a finite interval A, $s(A|u) = s(\bar{A}|u)$ will hold if $s(A|u) > -\infty$, and in case $s(A|u) = -\infty$ if $s(a',u) = s(a'',u) = -\infty$ also.

If $\left|\frac{U_o(q)}{|\Lambda|}\right| \leq K$ uniformly then the condition will hold at infinite end points $\pm\infty$.

<u>Proof:</u> $s(A \times \Delta) = \sup_{\substack{u_o \in A \\ u \in \Delta}} s(u_o,u) = \sup_{u \in \Delta} \sup_{u_o \in A} s(u_o,u) = \sup_{u \in \Delta} s(A|u)$. As a function

of Δ with A fixed $s(A \times \Delta)$ satisfies the regularity condition of Theorem 2, and hence by the uniqueness in Lemma 5 c) $s(A|u) = \inf_{\Delta \ni u} s(A \times \Delta)$

if $s(A|u)$ ·is u.s.c. at u . Similarly, $s(R|u) = \inf_{\Delta \ni u} s(R \times \Delta) = \inf_{\Delta \ni u} s(\Delta) =$

$= s(u)$ in this case.

To see that $s(A|u)$ is u.s.c. if $s(A|u) = s(\bar{A}|u)$ let

$s > s(A|u) = s(\bar{A}|u) = \sup_{u_o \in \bar{A}} s(u_o, u)$, so that $s(u_o, u) < s - \varepsilon$ for all

$u_o \in \bar{A}$ and some $\varepsilon > 0$. For any $u_o \in \bar{A}$, by the semicontinuity of

$s(u_o, u)$ in $\bar{R}^{M+1}$ there is then a neighbourhood $Au_o \times \Delta u_o \ni (u_o, u)$ such that $s(u'_o, u') < s - \varepsilon$ for all $(u'_o, u') \in Au_o \times \Delta u_o$ also. $\{Au_o ; u_o \in \bar{A}\}$ forms an open cover of $\bar{A}$, which is compact in $\bar{R}$. Hence a finite subfamily $\{Au_{o,i}\}_1^n$ already covers $\bar{A}$, so if $\Delta = \bigcap_1^n \Delta u_{o,i}$ = open neighbourhood of u then $s(u'_o, u') < s - \varepsilon$ for all $u'_o \in A$, $u' \in \Delta$, and $s(A|u') =$

$= \sup_{u_o \in A} s(u_o, u') \leq s - \varepsilon$ for $u' \in \Delta$, so $s(A|u)$ is u.s.c. at u.

If e.g. a' is a finite endpoint, and $s(A|u) > -\infty$, then $s(a', u) \leq s(A|u)$ by the convexity, because if $u_o \in A$ is such that $s(u_o, u) > s(A|u) - \varepsilon$, then $\dfrac{a' + u_o}{2} \in A$ too and hence

$$s(A|u) \geq s(\frac{a' + u_o}{2}, u) \geq \frac{1}{2} s(a', u) + \frac{1}{2} s(u_o, u) \geq \frac{1}{2} s(a', u) +$$

$$+ \frac{1}{2} s(A|u) - \frac{\varepsilon}{2} , \quad \text{so that}$$

$$s(a', u) \leq s(A|u) + \varepsilon \quad \text{for any} \quad \varepsilon > 0, \quad \text{and hence} \quad s(a', u) \leq s(A|u) .$$

If $\left| \dfrac{U_o(q)}{|\Lambda|} \right| \leq k$ then $s(A \times \Delta) = -\infty$ for all Δ if $A \subset (-\infty, k)$ or $A \subset (k, \infty)$, so $s(u_o, u) = \inf_{\substack{\Delta \ni u_o \\ \Delta \ni u}} s(A \times \Delta) = -\infty$ for all u if $k < |u_o| \leq \infty$.

Lemma 6 leads to the following important rule for how the probability distribution of $\dfrac{U_o(q)}{|\Lambda|}$ behaves in the thermodynamic limit: Consider the concave function $s(u_o, u)$ for u fixed. Typically the following picture is found:

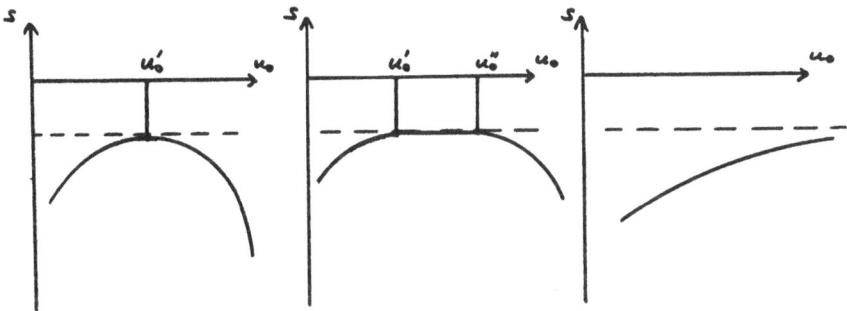

$s(u) = \sup_{u_0} s(u_0,u)$ is attained in an interval $M = [u_0',u_0'']$, which can reduce to a single point, finite or infinite. If A is an interval such that $\bar{A}$ is disjoint from M then

$s(A|u) = s(\bar{A}|u) < s(u)$, so

$$\lim_{\Lambda \to \infty} \frac{1}{|\Lambda|} \log P_\Lambda(A|\Delta) = s(A \times \Delta) - s(\Delta) = \sigma < 0$$

if $\Delta \ni u$ is small enough, i.e.

$$P_\Lambda(A|\Delta) \le e^{-|\Lambda|\sigma} \quad \text{as } \Lambda \to \infty ,$$

and the probability mass in A goes to zero very fast. This means that all probability mass will be concentrated to M in the limit. This is the basic rule in thermodynamic probability theory: The probability mass of a macroscopic variable $\dfrac{U_0(q)}{|\Lambda|}$ in the m. can. distribution defined by the restriction $\dfrac{U(q)}{|\Lambda|} \approx u$ will be concentrated to values of u_0 such that $s(u_0,u)$ is maximal. In particular, if $s(u_0,u)$ has a unique u_0' giving maximum then the law of large numbers holds: $\dfrac{U_0(q)}{|\Lambda|}$ converges to u_0' in probability. We are going to see that $s(u)$ (or rather $k \cdot s(u)$) is the (Boltzmann)-entropy of the macroscopic state $\dfrac{U(q)}{|\Lambda|} \approx u$, and the above argument makes precise the famous rules of thermodynamics that the probability of a macroscopic state $\dfrac{U_0(q)}{|\Lambda|} \approx u_0$ is proportional to

$\exp\left(\dfrac{\text{the entropy}}{k}\right) = e^{|\Lambda| s(u_o, u)}$, and that only states having maximal entropy are seen in macroscopic systems.

In the case where M is an interval $[u'_o, u''_o]$ the above argument does not tell us how the probability mass is distributed in M, only that the mass outside M goes to zero.

The corresponding result for the generalized canonical distribution defined on p. 41 follows analogously:

$$P_\Lambda\left(\dfrac{U(q)}{|\Lambda|} \in A \,\middle|\, a\right) = \dfrac{\Omega_\Lambda(A, a)}{\Omega_\Lambda(R^M, a)} \quad . \qquad \text{Hence}$$

$$\lim_{\Lambda \to \infty} \dfrac{1}{|\Lambda|} \log P_\Lambda(A|a) = s(A, a) - s(R^M, a)$$

(At least if both terms are not $+\infty$.)

Let us put $s(R^M, a) = g(a)$.

$s(A, a) = \sup_{u \in A} (s(u) - a \cdot u)$, and especially

$g(a) = \sup_{u} (s(u) - a \cdot u)$.

If M is the set $\subset R^M$ where $(s(u) - a \cdot u)$ is maximal it follows by the same argument as above that all the probability mass will be concentrated to M in the limit $\Lambda \to \infty$. Especially if M consists of a single point the law of large numbers holds for $\dfrac{U(q)}{|\Lambda|}$. The definition of M has a simple geometric interpretation: M is the set of u where a tangent plane of slope $a \in R^M$ touches the curve $s = s(u)$:

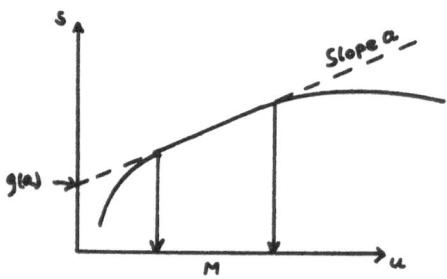

This means that M is always a convex set.

$g(a)$ is a convex function of a being the limit of the convex functions $g_\Lambda(a)$.

$\left(g_\Lambda(a) = \dfrac{1}{|\Lambda|} \log \int e^{-|\Lambda|(a \cdot u)} \Omega_\Lambda(du) \text{ is convex, because in any direction } b\right.$

its second derivative is ≥ 0: $\dfrac{d^2 g(a + \lambda b)}{d\lambda^2} \geq 0$.)

Since $s(A,a)$ is finite for any bounded A we see that if $g(a) = +\infty$ then the probability mass in any such A will go to zero, i.e. the mass escapes to infinity. This indicates that the system behaves in a "catastrophic" way; some quantities, perhaps the density will be infinite with probability close to one in a large system.

The relation between $g(a)$ and $s(u)$ is a well known conjugate relation in convexity theory:

Lemma 7: If $s(u)$ is concave and u.s.c. then its conjugate function $g(a) = \sup_u (s(u) - a \cdot u)$ is convex and lower semicontinuous (l.s.c.). If $g(a) < \infty$ for some a the relation is reciprocal in the sense that $s(u) = \inf_a (g(a) + a \cdot u)$, with a similar picture for the construction of s:

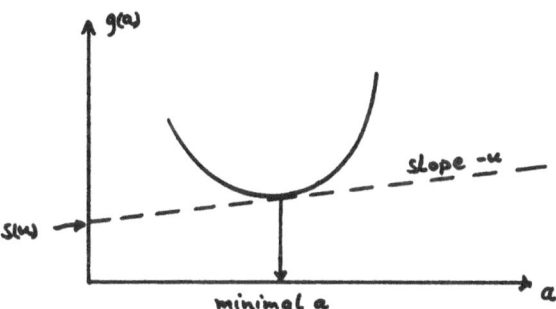

Two points (u,a) are related in the $\sup_u$ construction, i.e. $g(a) = s(u) - (a \cdot u)$ if and only if they are related in the $\inf_a$ construction, i.e. $s(u) = g(a) + (a \cdot u)$. There is a unique u corresponding to a if and only if g is differentiable at a, and then $u = - g'(a)$.

Proof: Consider first the degenerate case: $s(u) = -\infty$ for all u, then $g(a) = -\infty$ for all a, and $s(u) = \inf_a (g(a) + a \cdot u)$. $g(a)$ as defined above is the $\sup_u$ of the family of linear (and hence convex, l.s.c.) functions $g_u(a) = s(u) - a \cdot u$. Convexity and lower semicontinuity is preserved under arbitrary $\sup_u$ operations. Hence $g(a)$ has these properties.

Consider now the problem of finding all linear functions lying above s, and hence Epis:

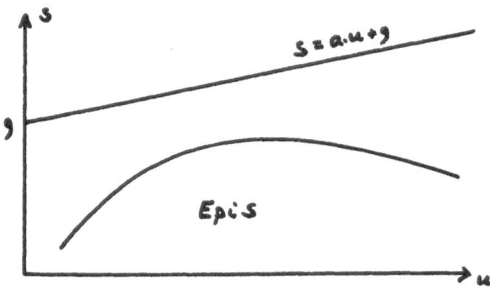

Such a function u → a·u + g lies above Epis iff. a·u + g ≥ s(u)
for all u, i.e. iff. g ≥ s(u) - a·u for all u, i.e. iff.
g ≥ sup s(u) - a·u = g(a). For a given a the "critical" g is hence
    u
g(a), when one tries to "push down" the function a·u + g as far as
possible towards Epis, and if g(a) = s(u) - a·u for some u then
a·u + g touches Epis at u. The possible values of (g,a) hence
correspond to Epig = {(g,a); g ≥ g(a)}. From g(a) ≥ s(u) - a·u for
all u,a follows that g(a) + a·u ≥ s(u) for all u,a , and hence
inf (g(a) + a·u) ≥ s(u). To see that there is indeed equality take any
 a
s > s(u). Then it is always possible to separate the point (s,u) from
Epis by a non vertical plane:

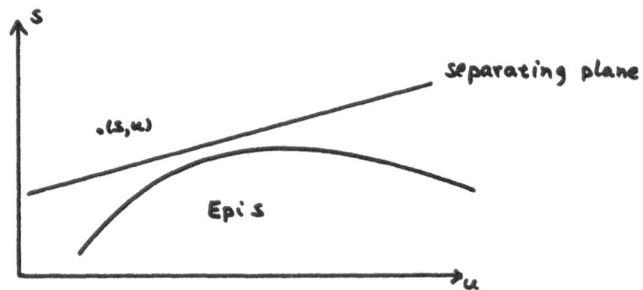

I.e. there is (g,a) such that a·u + g ≥ s(u) for all u but
a·u + g < s (Lemma 8 below). Hence g ≥ g(a), and
s > a·u + g(a) ≥ inf (g(a) + a·u), so in fact s(u) = inf (g(a) + a·u).
              a   a
Starting from the convex function g(a) and applying the above argument
we find that the linear function a → s - a·u lies below Epig iff.

s − a·u ≤ g(a) for all a i.e. iff. s ≤ inf (g(a) + a·u) = s(u) ,
                                              a
i.e. iff. (s,u) ∈ Epis:

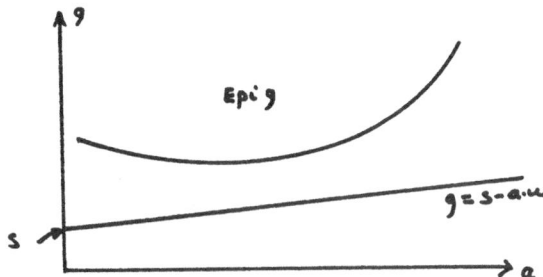

Hence the "critical" s for a given u is given by s = s(u), when one
tries to "push up" the function s − a·u as far as possible towards Epig.
We have hence seen that u → a·u + g touches Epis at u iff. g = g(a),
and a·u + g(a) = s(u) ; and that a → s − a·u touches Epig at a iff.
s = s(u), and s(u) − a·u = g(a). Hence these events are equivalent.

If u corresponds to a then

    g(b) ≥ s(u) − b·u   for all b   and
    g(a) = s(u) − a·u ,   so that
    g(b) − g(a) ≥ − u·(b − a)   for all b.

(−u defines a supporting plane to g at a.)

Hence it follows by letting b tend to a along any direction that if
the gradient g'(a) is defined then (b − a)·g'(a) = − (b − a)·u for all
(b − a) , so that u = − g'(a). Conversely, if g'(a) is not defined,
then there is a direction b such that the directional derivatives

$g'(a,b) = \lim\limits_{\lambda \to o} \dfrac{g(a + \lambda b) − g(a)}{\lambda}$ are different to the right and left:
$g'_+ \equiv g'(a,b) \neq − g'(a,-b) \equiv g'_-$. (These directional derivatives are
always defined, since g(a + λb) is convex in λ; and g'(a,b) ≥ − g'(a,-b).)
Both g'_+ and g'_- define supporting lines along b:

    $g(a + \lambda b) − g(a) \geq \lambda g'_\pm$ for all λ.

Hahn–Banach's theorem then tells us that there are two supporting planes
at a defined by $u_\pm$ which hence satisfy $g(a + c) − g(a) \geq c·u_\pm$ for
all c and which coincide with $g'_\pm$ along b, i.e. $b·u_\pm = g'_\pm$. Hence
both u_+ and u_- correspond to a and they are not equal, so there is
not a unique u corresponding to a.

For completeness we also give a proof of the separation property used above:

Lemma 8. Let s(u) be concave u.s.c. with g(a) < ∞ for some a , i.e. s(u) ≤ a·u + g for all u and some (g,a). Then if s > s(u) there exist (g,a) separating (s,u) from Epis, i.e. such that s(u) ≤ a·u + g for all u, and s > a·u + g .

Proof. We can assume that (s,u) = (0,0) by suitably moving the origin of the coordinates, so we assume that s(0) < 0 , and want to find a·u + g with g < 0 . The method of the proof is to consider the upward parabolas u → α|u|² + γ α > 0, γ arbitrary lying above Epis and try to push them towards Epis by minimizing γ :

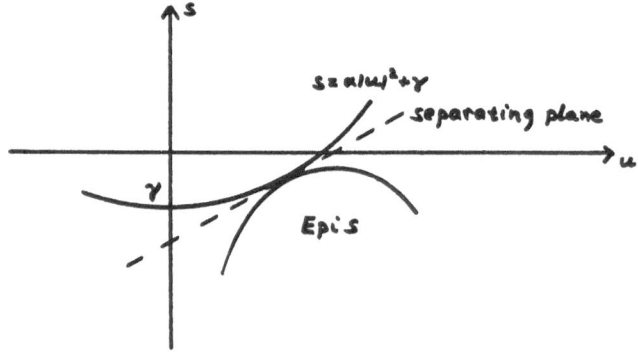

If the critical γ is < 0 then the tangent plane at the point of contact with Epis will be a separating plane. This will happen if α is big enough.

The parabola lies above Epis iff s(u) ≤ α|u|² + γ for all u , so the critical value of γ is given by γ = γ(α) = $\sup_u$ (s(u) - α|u|²) . To see that γ(α) < 0 if α is large enough we remember that the semicontinuity of s(u) at u = 0 means that s(u) ≤ - δ < 0 if |u| ≤ ε and ε is small enough. Hence

$$\gamma(\alpha) \leq \max\left(\sup_{|u|\leq\varepsilon}, \sup_{|u|>\varepsilon}\right) \leq$$

$$\leq \max\left(-\delta, \sup_{|u|>\varepsilon}(s(u) - \alpha\varepsilon|u|)\right) \leq$$

$$\leq \max\left(-\delta, \sup_{|u|>\varepsilon}|g| + (|a| - \alpha\varepsilon)|u|\right) =$$

$$= \max\left(-\delta, |g| + (|a| - \alpha\varepsilon)\varepsilon\right) = -\delta < 0$$

if α is large enough. Pick such a value of α and take $\gamma = \gamma(\alpha)$.

Consider $\gamma(\alpha) = \sup_u (s(u) - \alpha|u|^2)$. If it is $= -\infty$, $s(u) = -\infty$ for

all u, and we can take $a = 0$, g arbitrary < 0 to get the separating

plane. If it is $> -\infty$ it is actually finite and the sup is attained

for some u. This follows because $s(u) - \alpha|u|^2 \le |g| + |a||u| - \alpha|u|^2$,

which is bounded and goes to $-\infty$ when $|u| \to \infty$, so the $\sup_u = \sup_{|u| \le R}$

for R big enough, and an u.s.c. function always assumes a maximal

value on a compact set. Let $\hat{u}$ be the maximal u, so that

$s(u) - \alpha|u|^2 \le s(\hat{u}) - \alpha|\hat{u}|^2$ for all u, or

$s(u) - s(\hat{u}) \le \alpha|u|^2 - \alpha|\hat{u}|^2 = 2\alpha\hat{u}\cdot(u - \hat{u}) + \alpha|u - \hat{u}|^2$. It then follows

that actually $s(u) - s(\hat{u}) \le 2\alpha\hat{u}\cdot(u - \hat{u})$ for all u, because if for

some $u = \hat{u} + v$ we had $s(\hat{u} + v) - s(\hat{u}) \ge 2\alpha\hat{u}\cdot v + \delta$, $\delta > 0$, then by the

concavity of s near $\hat{u}$ when $u = \hat{u} + \varepsilon v = (1 - \varepsilon)\hat{u} + \varepsilon(\hat{u} + v)$ we would

have

$\quad s(u) \ge (1 - \varepsilon) s(\hat{u}) + \varepsilon\cdot s(\hat{u} + v) \ge s(\hat{u}) + (2\alpha\hat{u})\cdot(\varepsilon v) + \varepsilon\delta$

contradicting

$\quad s(u) \le s(\hat{u}) + (2\alpha\hat{u})\cdot(\varepsilon v) + \alpha\varepsilon^2|v|^2$

when ε is small enough, so that $\alpha\varepsilon^2|v|^2 < \varepsilon\cdot\delta$. We hence see that if

we take the tangent plane as $a\cdot u + g \equiv 2\alpha\hat{u}(u - \hat{u}) + s(\hat{u})$ it has desired

properties: $au + g \ge s(u)$ for all u, and $g = s(\hat{u}) - 2\alpha|\hat{u}|^2 =$

$= \gamma(\alpha) - |u|^2 < 0$, so it furnishes a solution to the problem.

The convex function $g(a)$ also allows us to study the limits of the

average values $\left\langle \frac{U(q)}{|\Lambda|} \right\rangle_{a,\Lambda}$ in the canonical ensemble:

For Λ finite we have $\left\langle \frac{U(q)}{|\Lambda|} \right\rangle_{a,\Lambda} = - g'_\Lambda(a)$. It then follows directly

from the convexity that $g'_\Lambda(a) \to g'(a)$ whenever the latter exists accor-

ding to the following lemma. In this case as we have seen $u = - g'(a)$

is the value to which $\frac{U(q)}{|\Lambda|}$ converges in probability as $\Lambda \to \infty$.

<u>Lemma 9</u>. If $g_\Lambda(a)$ are convex, differentiable and $g(a) = \lim_\Lambda g_\Lambda(a)$,

then $g(a)$ is convex, and $g'(a) = \lim_\Lambda g'_\Lambda(a)$ whenever $g'(a)$ exists.

<u>Proof</u>. The convexity of $g(a)$ is clear. If $g'(a)$ is defined then g is

finite in a neighbourhood of a. The convexity of $g_\Lambda(a)$ implies that

$g_\Lambda(a + b) - g_\Lambda(a) \ge b\cdot g'_\Lambda(a)$ for all b. Applying this for sufficiently

small $b = \pm \varepsilon e_i$ $i = 1,\ldots M$, ε fixed, such that $g_\Lambda(a + b)$ is finite
we find that $|\pm \varepsilon \cdot e_i \cdot g_\Lambda'(a)| \leq$ const as $\Lambda \to \infty$, so that $|g_\Lambda'(a)| \leq$ const.
Any limit point g' of the sequence $\{g_\Lambda'(a)\}$ will then satisfy the in-
equality $g(a + b) - g(a) \geq b \cdot g'$ for all b. But if $g'(a)$ exists this
means that $g' = g'(a)$ as we have seen, so all limit points are equal,
which means that the limit exists and is equal to $g'(a)$.

The relation between the m. can. and can. limiting values of typical observ-
ables $\dfrac{U_o(q)}{|\Lambda|}$ can now also be studied: The m. can. value of u_o in the
ensemble defined by u was that maximizing $s(u_o,u)$, supposing e.g. there
is only one such value. In the can. ensemble defined by a we include
$U_o(q)$ among $U(q)$ and put $a_o = 0$. Then

$$P_\Lambda \left(\dfrac{U_o(q)}{|\Lambda|} \in A \,\middle|\, a\right) = \dfrac{\Omega_\Lambda(A \times R^M,(0,a))}{\Omega_\Lambda(R^M,a)}, \quad \text{so as before}$$

$$\lim_{\Lambda \to \infty} \dfrac{1}{|\Lambda|} \log P_\Lambda = s(A \times R^M,(0,a)) - g(a) = \sup_{\substack{u_o \in A \\ u \in R^M}} (s(u_o,u) - a \cdot u) - g(a) .$$

Hence the probability mass will be concentrated to values of u_o such that

$$\sup_{u_o,u} (s(u_o,u) - a \cdot u) \text{ is attained.}$$

Supposing that u has a sharp value, i.e. that $g'(a)$ exists and
$u = - g'(a)$, we know that $\sup\limits_{u_o,u} (s(u_o,u) - a \cdot u) = \sup\limits_{u} (s(u) - a \cdot u) =$
$= s(u) - a \cdot u = \sup\limits_{u_o} s(u_o,u) - a \cdot u = s(u_o,u) - a \cdot u$, where u_o is the
m. can. value. Hence if u has a sharp value in the can. ensemble, then
the set M of maximal values of u_o in the two ensembles are the same.
Especially if u_o has a sharp value it is the same in the two ensembles.

Let us now collect the results about the law of large numbers in the different ensembles:

Theorem 3. Suppose that $s(u) > -\infty$ and the conditions of Lemma 6 are valid.
Then in the m. can. ensemble defined by $\dfrac{U(q)}{|\Lambda|} \approx u$ $P_\Lambda \left(\dfrac{U_o(q)}{|\Lambda|} \in M_o \,\middle|\, \Delta\right) \to 1$,
where M_o is the interval of values of u_o having maximal entropy $s(u_o,u)$.
Especially if there is a unique maximizing value the law of large numbers

holds for $\dfrac{U_o(q)}{|\Lambda|}$. In the can. ensemble defined by a, if $|g(a)| < \infty$

then $P_\Lambda(\dfrac{U(q)}{|\Lambda|} \in M|a) \to 1$, where M is the convex set in R^M where

$s(u) - (a \cdot u)$ is maximal, i.e. where the tangentplane $s = g(a) + (a \cdot u)$

of slope a touches Epis. This set reduces to a unique point u iff

$g'(a)$ exists, and then $u = - g'(a)$.

In this case $\left\langle \dfrac{U(q)}{|\Lambda|} \right\rangle_{\Lambda,a}$ also converges to u , and the set M_o for

any other variable $\dfrac{U_o(q)}{|\Lambda|}$ is the same for the two ensembles if u and

a are related by $u = - g'(a)$.

We have called $s(u)$ the m. can. entropy of the macroscopic state
defined by u. It can also be expressed as the entropy of the corresponding can. density as follows: The can. density is

$f(q) = \dfrac{e^{-a \cdot U(q)}}{G_\Lambda(a)}$, so if we define its (Gibbs)-entropy per unit volume by

$h_\Lambda(a) = \dfrac{-1}{|\Lambda|} \int f(q) \log f(q) \, \omega(dq)$, we see that it is equal to

$h_\Lambda(a) = \dfrac{1}{|\Lambda|} \log G_\Lambda(a) + \left\langle \dfrac{a \cdot U(q)}{|\Lambda|} \right\rangle_{\Lambda,a} = g_\Lambda(a) + \left\langle \dfrac{a \cdot U(q)}{|\Lambda|} \right\rangle_{\Lambda,a}$

Hence we see that if $g'(a)$ is defined, so there is a unique $u = - g'(a)$
corresponding to a then

$h(a) = \lim_{\Lambda \to \infty} h_\Lambda(a) = g(a) + (a \cdot u)$

with $u = - g'(a)$, i.e.

$h(a) = \inf_u (g(a) + a \cdot u) = s(u)$.

Hence $s(u)$ can also be computed as the can. Gibb´s entropy for a corresponding to u. Let us now return to the important special case to be treated
in the following, namely when $M = 2$ and U_1 is equal to the total energy H,
and U_2 is the total no. of particles. Then the parameters a_1 and a_2
will be $a_1 = \beta$, $a_2 = -\beta\mu$, and we will show that $s(e,n) > -\infty$ in an open
region D and differentiable there, and that it is an increasing function
of e. g is given by

$g(\beta,\mu) = \sup_{e,n} (s(e,n) - \beta e + \beta\mu n)$,

and will be shown to be finite when $\beta > 0$, μ arbitrary. Since $s(e,n)$ is

differentiable the values of e,n corresponding to β,μ will satisfy
the equations $\beta = s_e'(e,n)$, $\beta\mu = -s_n'(e,n)$, and then
$g(\beta,\mu) = s(e,n) - \beta e + \beta\mu n$. Conversely, if $g(\beta,\mu)$ is differentiable
the relation $s(e,n) = \inf_{\beta\mu} (g(\beta,\mu) + \beta e - \beta\mu n)$ imply that $e = - g_\beta'(\beta,\mu)$
(derivation with $\beta\mu$ = const.), $n = \beta^{-1}g_\mu'(\beta,\mu)$ for the values of e,n
corresponding to β,μ .

3.2. Derivation of the basic laws of thermodynamics
3.2.1. The rules for thermodynamic equilibrium

In thermodynamics one considers systems consisting of some parts in
Λ_1,Λ_2, etc. each of which when isolated can be described by e.g. a m. can.
law with a few macrovariables (e_1,n_1), (e_2,n_2) etc. The equilibrium
values of interesting variables are then determined by maximizing the
entropies of the isolated subsystems, subject to the constraints stipu-
lated for each of them. When these constraints are changed the values of
the macroscopic variables are changed to new equilibrium values, and one
is interested in the general rules for the directions of these changes.
Especially one wants to study how the energies are changed, what fraction
goes into heat, mechanical work etc.

Typical such changes: heat flows from a hot body to a cold one when an isolating wall is taken away, a gas in a cylinder expands when a piston is
allowed to move, the concentrations of various substances are changed when
they are allowed to mix, chemical reactions take place when the amounts of
different constituents are varied etc. To start with, let us illustrate the
arguments involved by considering three basic situations: thermal, pressure
and concentration equilibrium between two systems, and at the same time we
will define the basic intensive quantities temperature, pressure and chemical potential. Thermal equilibrium: Consider two systems in Λ_1,Λ_2 described by (e_1,n_1) and (e_2,n_2) which together form an isolated system,
and are separated by a wall, which can be changed from isolating to conduc-
ting with respect to exchange of energy:

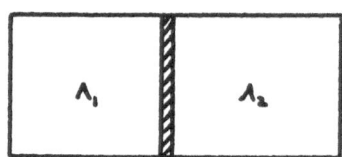

(In the following we usually denote the volume by V, $|\Lambda_1| = V_1$, $|\Lambda_2| = V_2$, $V = V_1 + V_2$, and volume fractions by v_i $v_1 = \frac{V_1}{V}$, $v_2 = \frac{V_2}{V}$.)

The total energy of a configuration $x = (x_1, x_2)$ is $H_1(x_1) + H_2(x_2)$, if the interaction with and across the wall can be neglected. (This interaction if present ought to be a surface effect of the magnitude $v^{2/3}$, and therefore small compared to the total energy of the magnitude V when $V \to \infty$.) This means that for the total system in $\Lambda = \Lambda_1 \cup \Lambda_2$ described by (e_1, n_1, e_2, n_2) we have

$$\Omega_\Lambda(A_1 \times A_2) = \Omega_{\Lambda_1}(A_1) \, \Omega_{\Lambda_2}(A_2) \qquad A_1, A_2 \subseteq R^2$$

(No permutation of particles between Λ_1 and Λ_2), and hence the entropies are additive:

$$\frac{1}{V} \log \Omega_\Lambda = v_1 \cdot \frac{1}{V_1} \log \Omega_{\Lambda_1} + v_2 \cdot \frac{1}{V_2} \log \Omega_{\Lambda_2} \,,$$

and in the limit $\Lambda_1, \Lambda_2 \to \infty$ v_1, v_2 = const. we get $s(A_1 \times A_2) = v_1 s_1(A_1) + v_2 s_2(A_2)$, and $s(e_1, n_1, e_2, n_2) = v_1 s_1 (\frac{e_1}{v_1}, \frac{n_1}{v_1}) + v_2 s_2 (\frac{e_2}{v_2}, \frac{n_2}{v_2})$.

$((\frac{H_1}{V}, \frac{N_1}{V}) \in A_1$ $(\frac{H_2}{V}, \frac{N_2}{V}) \in A_2$. are the restrictions defining $\Omega_\Lambda(A_1 \times A_2)$.)

If the wall permits the exchange of energy then the m. can. law for the system is defined by the restrictions

$$\frac{H_1 + H_2}{V} \approx e = e_1 + e_2 \qquad \frac{N_1}{V_1} \approx \frac{n_1}{v_1} \,, \quad \frac{N_2}{V_2} \approx \frac{n_2}{v_2} \,, \quad \text{so if } \Delta \subseteq R^4 \text{ is the set}$$

defined by these restrictions the distribution of e.g. $\frac{H_1}{V}$ is asymptotically given by:

$$P_\Lambda \, (\frac{H_1}{V} \in A | \Delta) = \sup_{\Delta \cap \{e_1 \in A\}} s(e_1, n_1, e_2, n_2) - \sup_\Delta s(e_1, n_1, e_2, n_2) \,,$$

and in the limit when Δ shrinks we see that the values taken by $\frac{H_1}{V}$ are those determined by $\sup_{e_1} s(e_1, n_1, e_2, n_2)$ when

$$e_1 + e_2 = e \,, \quad \text{i.e. by}$$

$$\sup_{e_1} (v_1 s_1 \, (\frac{e_1}{v_1}, \frac{n_1}{v_1}) + v_2 s_2 \, (\frac{e - e_1}{v_2}, \frac{n_2}{v_2})) \,.$$

The equilibrium values of e_1 hence have to satisfy the equation

$$\frac{\partial s_1(e_1,n_1)}{\partial e_1} = \frac{\partial s_2(e_2,n_2)}{\partial e_2} \quad , \quad \text{i.e.} \quad \beta_1 = \beta_2$$

This makes it natural to identify $\beta = \dfrac{\partial s(e,n)}{\partial e}$ as the (inverse) temperature of a system, because then the equilibrium condition for thermal
contact is that the temperatures of the subsystems are equal, and for an
ideal gas it coincides with our ordinary scale of temperature. From the
convexity of s_1,s_2 follows the important fact that $\beta = \dfrac{\partial s(e,n)}{\partial e}$ is a decreasing function of e. This means that that system which gives away
energy to the other is the one which increases its β, i.e. the one
which decreases its temperature $(= \frac{1}{\beta})$. We hence see that the convexity
of s is intimately related with our most elementary physical experiences.
The fact that $s(e,n)$ is increasing in e means that β always has to
be ≥ 0. (If a system had β negative it would always give away energy in
thermal contact with another system however hot, a very weird property. It
would be hotter than any other system with $\beta \geq 0$.)

Pressure equilibrium: Consider now the situation when the wall is conduc-
ting and also free to move as a piston not allowing exchange of particles.

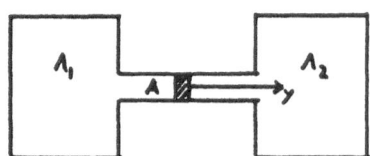

Include its position y or equivalently v_1 as a state variable $(\dfrac{dv_1}{dy} = \dfrac{A}{V}$,
where A is the area of the piston). Ω_Λ for a state defined by

$$\frac{H_1}{V} \approx e_1 \quad , \quad \frac{H_2}{V} \approx e_2 \quad , \quad \frac{N_1}{V} \approx n_1 \quad , \quad \frac{N_2}{V} \approx n_2 \quad , \quad y \approx \text{fixed}$$

is hence:

$$\Omega_\Lambda = \Omega_{\Lambda_1} \left(\frac{de_1}{v_1}, \frac{dn_1}{v_1}\right) \Omega_{\Lambda_2} \left(\frac{de_2}{v_2}, \frac{dn_2}{v_2}\right) dy \quad , \quad \text{so}$$

$$s(e_1,n_1,e_2,n_2,y) = v_1 s_1\left(\frac{e_1}{v_1}, \frac{n_1}{v_1}\right) + v_2 s_2\left(\frac{e_2}{v_2}, \frac{n_2}{v_2}\right)$$

$(\frac{1}{V} \log \Delta y = \frac{1}{V} \log \frac{V}{A} \Delta v_1 \to 0$ as $V \to \infty$ even if $A \sim V^{2/3}$.)

Hence by the same argument as above the equilibrium values of y or equivalently v_1, v_2 and e_1, e_2 are such that s is maximal, i.e. the condition is:

$$\sup v_1 s_1 \left(\frac{e_1}{v_1}, \frac{n_1}{v_1}\right) + v_2 s_2 \left(\frac{e_2}{v_2}, \frac{n_2}{v_2}\right)$$

when $v_1 + v_2 = 1$, $e_1 + e_2 = e = $ fixed, n_1, n_2 fixed.

The equ. condition for e_1, e_2 gives as before $\beta_1 = \dfrac{\partial s_1}{\partial e_1} = \beta_2 = \dfrac{\partial s_2}{\partial e_2}$, and for v_1, v_2 we get:

$$\left(s_1 - \frac{e_1}{v_1} \frac{\partial s_1}{\partial e_1} - \frac{n_1}{v_1} \frac{\partial s_1}{\partial n_1}\right) dv_1 + \left(s_2 - \frac{e_2}{v_2} \frac{\partial s_2}{\partial e_2} - \frac{n_2}{v_2} \frac{\partial s_2}{\partial n_2}\right) dv_2 = 0$$

when $dv_1 + dv_2 = 0$.

I.e. with $\beta = \dfrac{\partial s}{\partial e}$, $\beta\mu = -\dfrac{\partial s}{\partial n}$

$$\left(s_1 - \beta_1 \frac{e_1}{v_1} + \beta_1 \mu_1 \frac{n_1}{v_1}\right) = \left(s_2 - \beta_2 \frac{e_2}{v_2} + \beta_2 \mu_2 \frac{n_2}{v_2}\right).$$

But when $\beta = \dfrac{\partial s}{\partial e}$ $\beta\mu = -\dfrac{\partial s}{\partial n}$ we have $g(\beta, \mu) = s(e, n) - \beta e + \beta\mu n$, so the equilibrium condition is

$$\beta_1 = \beta_2, \quad g_1 = g_2, \quad \text{or}$$

$$\beta_1 = \beta_2, \quad \beta_1^{-1} g_1 = \beta_2^{-1} g_2$$

Earlier we saw that $\beta^{-1} g = p = $ the pressure, so we arrive at the equilibrium conditions $\beta_1 = \beta_2$, $p_1 = p_2$, for thermal and pressure equilibrium. The identification of p can also be seen directly if we replace the system Λ_2 by a constant force pA on the piston. The energy of Λ_1 is then $H_1 + pAy$, so the equ. values of e_1, v_1 are determined by

$$\sup v_1 s_1 \left(\frac{e_1}{v_1}, \frac{n_1}{v_1}\right) \quad \text{when}$$

$$e_1 + pay = e = \text{const}, \quad v_1 + v_2 = 1 \quad a = \frac{A}{V}$$

Hence the equ. conditions are

$$g_1 dv_1 + \beta de_1 = 0, \quad \text{when}$$

$de_1 + pady = 0$, $dv_1 = ady$, i.e. $(g_1 - \beta_1 p_1) dv_1 = 0$, and we see that $\beta p = g$ as before.

Also in this case the convexity has a very basic physical interpretation:

For β fixed $g(\beta,\mu)$ is convex and increasing in μ, hence $\frac{\partial g}{\partial \mu}$ is also increasing in μ, but $\frac{\partial g}{\partial \mu} = \beta n$, so we see that if μ is eliminated $\beta p = g$ is an increasing function of n. Hence if V is varied with N constant, so $n = \frac{\bar{n}}{v}$, we see that p decreases as v increases, as we expect.

Concentration equilibrium: If in the above situation n_1 and n_2 are also free to vary subject to $n_1 + n_2 = n = $ const. then we must also have $\frac{\partial s_1}{\partial n_1} = \frac{\partial s_2}{\partial n_2}$ for the equ. values, i.e. $\beta_1 \mu_1 = \beta_2 \mu_2$, or $\mu_1 = \mu_2$.

(The above definitions of β and p have the advantage that they do not suppose that there is a unique value of e.g. e corresponding to a given β. If not $s(e,n)$ is linear with slope β on an interval and any e in this interval gives $\frac{\partial s(e,n)}{\partial e} = \beta$. This will happen at phase transitions.)

The methods illustrated above show the structure of the rules of thermo-dynamics which determine the values of macrovariables in equilibrium: A closed system consists of a number of weakly interacting parts $\Lambda_1, \Lambda_2, \ldots$ each described by a few macrovariables $U_1, U_2, \ldots$ having entropy functions $s_1(u_1), s_2(u_2), \ldots$ The entropy of the state of the whole system defined by $\frac{U_1}{V} \approx u_1$, $\frac{U_2}{V} \approx u_2$, $\ldots$ is then

$$s(u_1, u_2 \ldots) = v_1 s(\frac{u_1}{v_1}) + v_2 s_2(\frac{u_2}{v_2}) + \ldots$$

$\frac{V_i}{V} = v_i$, where V is a macroscopic volume, not necessarily $= \sum_i V_i$. The probability distribution of $\frac{U_1}{V}, \frac{U_2}{V}, \ldots$ in a m. can. ensemble defined by a number of additive restrictions, which can be written in general as

$$\sum_i D_i u_i + d_i v_i = u$$

with given matrices D_i and vectors d_i defining the couplings between the variables u_i and v_i respectively, is asymptotically given by

$\exp V\left[\sum_i v_i s_i(\frac{u_i}{v_i}) - s(u)\right]$, where $s(u) = \sup_{u_i, v_i} \sum_i v_i s_i(\frac{u_i}{v_i})$ under the above restrictions. Hence the equ. values of u_i, v_i which get probability one are those which give the sup in $s(u)$. Hence when new restrictions are imposed the system will move to a state such that $\sum_i v_i s_i(\frac{u_i}{v_i})$ becomes maximal under these new restrictions; any change from one to another equ. position that takes place under given restrictions will be such that the total entropy $\sum_i v_i s_i(\frac{u_i}{v_i})$ increases to its maximal value $s(u)$. "Die Energie der Weld bleibt konstant, und die Entropie strebt einem Maximum zu." The equ. values of u_i, v_i satisfy equ. conditions which mean that various intensive variables balance each other. The general structure of these equations can be seen as follows:

Consider also the ensemble where the restrictions are relaxed, and a parameter (row-) vector a is introduced corresponding to u. Then $g(a)$ is given by:

$$g(a) = \sup_u \ (s(u) - a \cdot u) =$$

$$= \sup_{u_i, v_i} \ \sum_i v_i (s_i(\frac{u_i}{v_i}) - a \cdot D_i \frac{u_i}{v_i}) - a \cdot d_i) =$$

$$= \sup_{v_i} \ \sum_i v_i (\sup_{u_i} (s_i(u_i) - a \cdot D_i u_i) - a \cdot d_i) =$$

$$= \sup_{v_i} \ \sum_i v_i (g_i(a \cdot D_i) - a \cdot d_i) \ .$$

Now, it is easy to see that $s(u)$ is concave in u (Lemma 10 below). Then we can use the basic facts about duality in Lemma 7 to $s(u)$ and $g(a)$, and we find:

If we have $a_i = a \cdot D_i$ such that $g_i(a_i) = a \cdot d_i$ then $g(a)$ is finite $= 0$, in fact $\sum_i v_i(g_i(a_i) - a \cdot d_i) = 0$ for all v_i. Also, if $g_i(a_i) =$

$$= s_i(\frac{u_i}{v_i}) - (a_i \cdot \frac{u_i}{v_i}) \ , \quad \text{i.e. if} \quad a_i = s_i'(\frac{u_i}{v_i}) \ , \quad \text{and if } u_i, v_i \text{ satisfy the}$$

restrictions then they give the $\sup\limits_u$ because

$$g(a) = \sum_i v_i (s_i(\frac{u_i}{v_i}) - a \cdot D_i \frac{u_i}{v_i} - a \cdot d_i) =$$

$$= \sum_i v_i s_i(\frac{u_i}{v_i}) - a \cdot u \ , \quad \text{and hence}$$

$$\sum_i v_i s_i(\frac{u_i}{v_i}) = s(u) \ , \quad \text{because} \quad g(a) \geq s(u) - a \cdot u \text{ always.}$$

Conversely, if we have u_i, v_i such that they satisfy the restrictions and

$\sum_i v_i s_i (\frac{u_i}{v_i}) = s(u)$, then if there exists a nonvertical tangent plane

to $s(u)$ at u then there is a such that $g(a)$ is finite and

$g(a) = s(u) - (a \cdot u)$. Then (unless some $v_i = 0$) $\quad g(a \cdot D_i) = a \cdot d_i$ for

all i, and putting $a_i = a \cdot D_i$

$0 = g(a) - s(u) + (a \cdot u) =$

$$= \sum_i v_i (g_i(a_i) - (a \cdot d_i) - s_i(\frac{u_i}{v_i}) + (a_i \frac{u_i}{v_i}) + (a \cdot d_i)) =$$

$$= \sum_i v_i (g_i(a_i) - s_i(\frac{u_i}{v_i}) + (a_i \frac{u_i}{v_i})) .$$

But then $g_i(a_i) = s_i(\frac{u_i}{v_i}) - a_i \frac{u_i}{v_i}$ for all i, because

$g_i(a_i) \geq s_i \frac{u_i}{v_i} - a_i \frac{u_i}{v_i}$ always. This means that a_i and $\frac{u_i}{v_i}$ are

related by $a_i = s_i'(\frac{u_i}{v_i})$. We also note that if $s(u)$ is differentiable

at u then $a = s'(u)$. Let us collect these results:

Theorem 4. Consider a system composed of weakly interacting parts as
described above. Suppose that $s(u)$ defined by $s(u) = \sup \sum_i v_i s_i (\frac{u_i}{v_i})$
when

$$\sum_i D_i u_i + d_i v_i = u , \quad v_i \geq 0$$

is u.s.c. and that it has a non vertical tangent plane at u. Then
$\{u_i, v_i\}$ satisfying the restrictions and having $v_i > 0$ are equilibrium
values, i.e. give $\sup_u$ in $s(u)$ if and only if the corresponding intensive parameters $a_i = s_i'(\frac{u_i}{v_i})$ satisfy the balance equations:

$$a_i = a \cdot D_i , \quad g_i(a_i) = a \cdot d_i ,$$

for some value a of the parameter vector. If $s(u)$ is differentiable
at u then $s'(u) = a$. The equations $a_i = a \cdot D_i$ express thermal etc.
equilibrium, and $g_i(a_i) = a \cdot d_i$ pressure equilibrium. $s(u)$ is homogenous:

$$s(\alpha u) = \alpha s(u) \quad \text{for} \quad \alpha > 0 .$$

We also give the proof of the convexity of $v_i \cdot s(\frac{u_i}{v_i})$ and of $s(u)$.

Lemma 10. If $s(u)$ is any concave u.s.c. function then $v \cdot s(\frac{u}{v})$ has the same properties as a function of (u,v). Also $s(u)$ defined above is concave.

Proof. If $(u,v) = \lambda(u',v') + (1 - \lambda)(u'',v'')$ then

$$\frac{u}{v} = (\frac{\lambda v'}{v}) \frac{u'}{v'} + (\frac{(1 - \lambda)v''}{v}) \frac{u''}{v''} , \quad \text{so}$$

$$s(\frac{u}{v}) \geq \frac{\lambda v'}{v} s(\frac{u'}{v'}) + \frac{(1 - \lambda)v''}{v} s(\frac{u''}{v''}) , \quad \text{or}$$

$$vs(\frac{u}{v}) \geq \lambda v' s(\frac{u'}{v'}) + (1 - \lambda)v'' s(\frac{u''}{v''}) .$$

If $s - \epsilon > vs(\frac{u}{v})$ then this is still true if u varies slightly because s is u.s.c. Hence it is also true if both u and v vary slightly.

To see that $s(u) \geq \lambda s(u') + (1 - \lambda)s(u'')$ if $u = \lambda u' + (1 - \lambda)u''$ with $s(u)$ defined above take for any $\epsilon > 0$ values (u_i',v_i') (u_i'',v_i'') correspon-ding to u',u'' such that

$$\sum_i v_i' s_i (\frac{u_i'}{v_i'}) > s(u') - \epsilon \quad \text{etc.}$$

Then $(u_i,v_i) = \lambda(u_i',v_i') + (1 - \lambda)(u_i'',v_i'')$ correspond to u because the restrictions are linear, and

$$s(u) \geq \sum_i v_i s_i (\frac{u_i}{v_i}) \geq$$

$$\geq \lambda \sum_i v_i' s_i (\frac{u_i'}{v_i'}) + (1 - \lambda) \sum_i v_i'' s_i (\frac{u_i''}{v_i''}) \geq$$

$$\geq \lambda s(u') + (1 - \lambda)s(u'') - \epsilon .$$

Since this is true for any $\epsilon > 0$ $s(u)$ is convex.

Remark: If there is only one system the above situation corresponds to having the restrictions $\frac{U_1}{V} \approx u$, $\frac{V_1}{V} = v$ and $V \to \infty$. Then

$s(u,v) = v \cdot s_1(\frac{u}{v})$. This is the convenient way of scaling the variables when one wants to consider a system whose volume V_1 as well as U_1 can be varied.

If moreover $u = (e,n)$, so $s(u,v) = vs(\frac{e}{v},\frac{n}{v})$, then the components of $a = s'(u,v)$ are $a^1 = s_e' = \beta$, $a^2 = s_n' = - \beta \cdot \mu$ and

$\overset{3}{a} = s - \frac{e}{v} s'_e - \frac{n}{v} s'_n = g = \beta \cdot p$, so in an infinitesimal change we have

$ds(e,n,v) = \beta de - \beta \mu dn + \beta p dv.$

The equation for a^3 can also be written

$s(e,n,v) = \beta e - \beta \mu n + \beta p v$,

which together with the previous one expresses the fact that $s(e,n,v)$ is homogenous (Euler's theorem).

We can now prove a version of the statement that the probability law for a small subsystem Λ_1 in contact with a big system Λ_2 (heat bath) is given by the can. law:

The equ. values of u_1 and u_2 are determined by

$\underset{u_1,u_2}{\sup} \ v_1 s_1 \left(\frac{u_1}{v_1}\right) + v_2 s_2 \left(\frac{u_2}{v_2}\right)$ when

$u_1 + u_2 = u = $ fixed

$v_1, v_2 = $ fixed .

'he equ. equations are hence

$s'_1 \left(\frac{u_1}{v_1}\right) = s'_2 \left(\frac{u_2}{v_2}\right) \equiv a$

If we now consider the situation when $v_2 \to \infty$, v_1 fixed, $\frac{u}{v_2} \to \bar{u} = $ fixed, we see that $\frac{u_2}{v_2} = \bar{u} - \frac{u_1}{v_2} \to \bar{u}$ when u_1 remains bounded. Hence if we consider any limiting value of u_1, $u_1 \to \bar{u}_1$ we see that it has to satisfy the equ. equation $s'_1 \left(\frac{\bar{u}_1}{v_1}\right) = s'_2 (\bar{u}) \equiv a$ if s'_i are continuous. This is now the case by Lemma 9: If $s(u)$ is concave and differentiable and $u_n \to \bar{u}$ then $s_n(u) \equiv s(u + u_n) \to \bar{s}(u) \equiv s(u + \bar{u})$, because $s(u)$ is continuous if it is differentiable. Hence by Lemma 9 $s'_n(u) \to \bar{s}'(u)$, i.e. $s'(u + u_n) \to s'(u + \bar{u})$, and especially $s'(u_n) \to s'(\bar{u})$, i.e. $s'(u)$ is continuous. In Theorem 3 we saw that the values of u_1 in the can. ensemble determined by a were those where $s'_1 \left(\frac{u_1}{v_1}\right) = a$. Hence we see that any limit value of u_1 is an equ. value in the can. ensemble determined by $a = s'_2(\bar{u})$. Especially, if there is a unique such value we see that u_1 converges to it as $v_2 \to \infty$:

<u>Theorem 5</u>. Any equ. value of u_1 in a small subsystem Λ_1 in contact with a large system Λ_2 will be an equ. value in the can. ensemble for Λ_1 determined by $a = s_2'(\bar{u})$ in the limit $u_1 + u_2 = u$, v_1 fixed, $v_2 \to \infty$, $\frac{u}{v_2} \to \bar{u}$. Especially, if a determines u_1 uniquely then u_1 will converge to this unique value as $v_2 \to \infty$.

3.2.2. The rules for energy changes; work, heat and their relation to entropy

With the help of the general rules derived in the previous section we can now study how the energies of the subsystems change when the equilibrium values of the state variables change. In this process we will see how to distinguish between that part of the energy which is called heat and that part which is called (useful) work. We will see how the celebrated rules for the operation of machines converting heat and work into each other come out of the laws for the equilibria.

Consider first the definition of work. A typical system where energy can be stored and be made useful at a later time is a suspended weight:

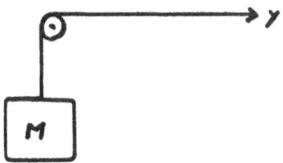

Its energy is a function of y: $E = Mgy + $ (internal energy), and by raising it potential energy is stored, which can be retrieved when needed by lowering it. Its entropy depends only on the internal energy and is hence constant when $E = Mgy + E_o$. (Neglecting energy lost by friction in the wheel etc.) This is the property that makes it useful for storing energy, because when coupled to a similar system e.g.:

the equilibrium positions of y in the total system are such that the
total entropy is maximal under the condition that the total energy is
fixed. Since the entropy is independent of y it is not determined by
the maximization, and by an infinitesimal perturbation (kick) it can
be moved a large amount back and forth. Then the other system can be dis-
connected and work has been taken in or out essentially without loss.
Similarly, if it is coupled to the system with the piston considered before:

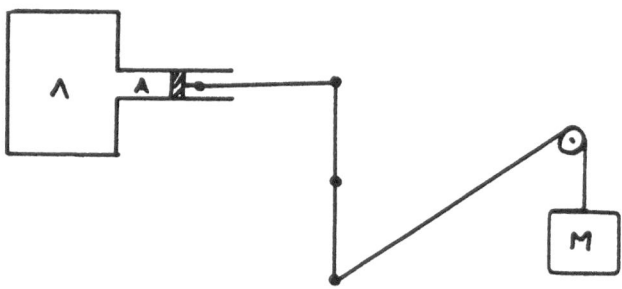

then as we saw the equ. position is determined by $\sup_{v}$ vs $(\frac{e}{v}, \frac{n}{v})$ when

$v - ay = v_o$ = const.

because there is no contribution to the entropy from the weight, and we
assume that Λ and the weight are enclosed by isolating walls. If we
design the machine carefully by making M a function of y we can how-
ever make vs$(\frac{e}{v}, \frac{n}{v})$ independent of y . The condition for this to happen
is that $s'_e de + g dv = 0$, or de = - pdv. If M is changed by an amount
dM which is chopped off and left at the height where it is located then
no work is done in this process, and the energy of the whole system including
the mass chopped off is constant, i.e. de + Mgdy = 0. Hence, if M is
varied so that $\frac{Mg}{a}$ = p all the time when v is varied then vs $(\frac{e}{v}, \frac{n}{v})$ will
be independent of v, and as before it can be varied back and forth by an
infinitesimal effort, so that work can be taken in or out of Λ and be
stored in the weight.

In the above discussion we have considered infinitesimal changes where the
macrovariables all the time take their equilibrium values determined by the
(variable) restrictions. Such changes are called quasistatic, and take place
if the rate of change of the restrictions is slow compared to the time it
takes for the variables to reach their equilibrium values with constant
restrictions. Such a change in which the total entropy of the system is
constant is called reversible. Any change in a closed system with given

restrictions which can also go backwards must be reversible in this sense, because in both the entropy change is ≥ 0; if $\Delta s \geq 0$ and $-\Delta s \geq 0$ then $\Delta s = 0$. In general a reversible change can be made to run back or forth by an infinitesimal extra effort as above. A general system of the type considered before whose thermodynamic state is determined by a (vector-) variable u and has entropy $s(u)$ can be called a work source for some set of possible quasistatic variations of u if its entropy is constant under these variations. Correspondingly, the change in the energy of a system Λ_1 which occurs in a reversible change of its state is called work, because if the system is coupled to a work source Λ_2 so that

$$\sum_1^2 D_i u_i + d_i v_i = u = \text{const.}$$

and energy is exchanged between the systems then the work can be transferred to Λ_2 in a reversible change of the total system with an infinitesimal effort as in the example. In an infinitesimal part of a quasistatic change of a system we have:

$$ds = a \cdot du \quad \text{since} \quad s'(u) = a \ .$$

Hence if the components of u are $(e, u^2, \ldots u^L)$, and those of a are called $(\beta, -\beta\alpha^2, \ldots -\beta\alpha^L)$ then for a reversible change we have $ds = 0$, or $\beta de - \beta \sum_2^L \alpha^i du^i = 0$. Hence the infinitesimal work put into the system is given by $\delta w = \sum_2^L \alpha^i du^i$, and the total work put in is the integral of this expression along the quasistatic path describing the change. When $u = (e, n, v)$, $\delta w = \mu dn - p dv$ (p. 66) . We now come to the definition of heat. A typical heat source is a gas in a container with constant volume, which only changes its state by exchange of energy quasistatically through thermal interaction with other systems. The energy put into it in such a change is called heat. If the system is of the general type we have considered we have $ds = \beta de$ if only e and not $u^2, \ldots u^L$ change. Hence the heat added in an infinitesimal quasistatic change of a heat source is given by $\delta q = \beta^{-1} ds = kT ds$ when $du^i = 0$, $i = 2, \ldots L$. If an arbitrary system Λ_1 is coupled to a heat source and only the energies are allowed to vary subject to $e_1 + e_2 = e$ then the energy change in the system will be called heat. In an infinitesimal quasistatic change we also have $ds_1 = \beta_1 de_1$ when only e_1, e_2 change, so also for an arbitrary system $\delta q = \beta^{-1} ds$ when only the energy changes, and not the other variables.

If we consider an arbitrary infinitesimal quasistatic change in a system

with $ds = a \cdot du = \beta de - \beta \sum\limits_{2}^{L} \alpha^i du^i$ then de can be split into

$de = \delta w + \delta q$ with $\delta w = \sum\limits_{2}^{L} \alpha^i du^i$ and

$\delta q = \beta^{-1} ds$.

Correspondingly the change can be achieved in two steps: First a reversible change when ds stays constant, then an irreversible change of e with $du^2, \ldots du^L = 0$.

The first change can be thought of as brought about by the interaction with a suitable work source, and the second one by the interaction with a heat source, and in this process δw and δq are the work and heat added to the system.

Hence we have arrived at the two fundamental laws for quasistatic changes in thermodynamics

$de = \delta w + \delta q$ (first law)

$\delta q = (kT)ds$ (second law)

The total change in e and s along a quasistatic path between two states u' and u'' is hence given by

$e'' - e' = \int\limits_{u'}^{u''} \delta w + \delta q$, and

$s'' - s' = \int\limits_{u'}^{u''} \frac{\delta q}{kT}$

Here the integrals of δw and δq depend on which path is chosen from u' to u'' , but $e'' - e'$, and $s'' - s'$ are independent of the path since e and s are functions of $u = (e, u^2, \ldots u^L)$.

This property is the important one in the treatment of thermodynamic problems.

In a given change $\delta w, \delta q$ and kT can in principle be measured empirically, and then the second law shows how s can be found empirically.

(Hence thermodynamics is the science where probabilities can be measured with thermometers and calorimeters.)

Let us now see how these concepts can be used to analyse the possibilities
of converting heat into work and vice versa. Consider a system in Λ (e.g.
a steam engine of some sort) which is coupled to two heat sources and
a work source:

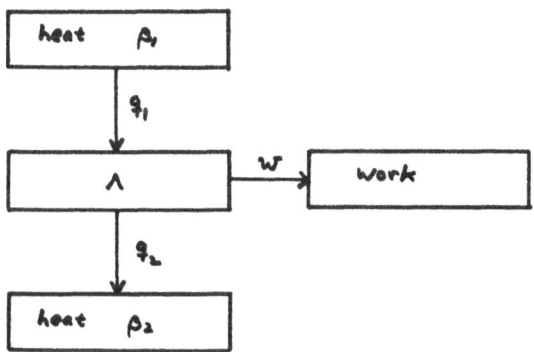

The combined system is isolated, and for simplicity assume that the heat
sources are so big that their temperatures do not change much when heat
is taken in or out. Suppose that the machine can run in a cycle thereby
taking an amount of heat q_1 and delivering part of it as work w to
the work source and part of it as lost heat q_2. We want to know what is
the maximal w that can be obtained. Consider one cycle of the operation.
Since the state of Λ returns to its initial value its energy and entropy
does not change. The work source increases its energy by w, and its
entropy does not change. The changes in the heat baths are $\Delta e_1 = - q_1$,
$\Delta s_1 = - \beta_1 q_1$, $\Delta e_2 = q_2$, $\Delta s_2 = \beta_2 q_2$ respectively.

(Assuming that the changes in them are quasistatic.) Since the combined
system is closed we must have $\Delta e = 0$ and $\Delta s \geq 0$, since the motion is
from one equ. state to another. (We can imagine an isolating wall enclosing
Λ opened at the beginning of the cycle and closed just when Λ returns to
its initial state, this state being an equ. state of Λ when it is isolated.)

Hence we must have

$\quad q_1 - q_2 - w = 0$, and

$\quad \beta_2 q_2 - \beta_1 q_1 \geq 0$.

This means that if $w > 0$ then $q_1 > q_2$, and $\beta_2 q_2 \geq \beta_1 q_2$, i.e. $\beta_2 > \beta_1$.

(We can not have $q_2 = 0$ because then $\beta_1 q_1 \leq 0$ contradicting $q_1 > q_2 = 0$.)

Hence we arrive at the fundamental conclusion that we must have $T_1 > T_2$ if the machine is to deliver work. This is the famous statement that a perpetum mobile of the second kind is impossible. Such a machine would deliver a positive w with $T_1 \leq T_2$. We also see that $w = q_1 - q_2 \leq$

$$\leq q_1 - \frac{\beta_1}{\beta_2} q_1 = q_1 (1 - \frac{\beta_1}{\beta_2})$$ with equality iff $\Delta s = 0$, i.e. if the whole

system moves reversibly during the cycle. This important conclusion means

that the efficiency $\frac{w}{q_1} \leq (1 - \frac{T_2}{T_1})$, and that the maximal efficiency is

achieved if the system works reversibly. Also, if $w < 0$, $q_1 < 0$, we

see that $\left|\frac{w}{q_1}\right| \geq (1 - \frac{T_2}{T_1})$. This means that if heat is to be transported

from a cold to a hot reservoir it is necessary to spend work, and the

efficiency is $\left|\frac{q_1}{w}\right| \leq \dfrac{1}{1 - \dfrac{T_2}{T_1}}$.

The interest in these conclusions is that they assume very little about the structure of the systems and that the efficiencies are universal functions of the temperatures and do not depend on the systems used.

Similarly, we can consider a system in Λ whose state $u = (e, u^2, \ldots u^L)$ changes by Δu from a given initial to a given final state thereby delivering a work w to a work source, and changing the state of the environment by $-(\Delta e + w, \Delta u^2, \ldots \Delta u^L)$. Again, the whole system is closed and we assume that $a = (\beta, -\beta \alpha^2, \ldots -\beta \alpha^L)$, the intensive parameters of the environment do not change much in the process, so its entropy changes by $-\beta(\Delta e + w) + \beta \sum_2^L \alpha^i \Delta u^i$.

If the process is a possible motion then we have:

$\Delta s - \beta(\Delta e + w) + \beta \sum_2^L \alpha^i \Delta u^i \geq 0$, with equality if the process is reversible. Hence

$$w \leq \beta^{-1} \Delta s - \Delta e + \sum_{2}^{L} \alpha^i \Delta u^i = w_{rev} \,, \quad \text{and again we see that the maximal}$$

work is obtained in a reversible change. We also see that if the change is cyclic so that $\Delta u = \Delta s = 0$, then $w \leq 0$, so no cyclic machine taking in energy from one fixed environment and delivering it as work can be constructed. (The function $-\beta^{-1} s + e - \sum_{2}^{L} \alpha^i u^i =$

$= -\beta^{-1}(s - a \cdot u)$ measures the useful work that can be obtained from the system in state u in an environment with parameters a. It has been called exergy and is nowadays much used in discussions of energy re-sources. It takes its minimal value when u is the value which is in equilibrium with the environment.

When $u = (e,n,v)$, $a = (\beta,-\beta\mu,\beta p)$ e.g. we have

$$-\beta^{-1}(s - a \cdot u) = -\beta^{-1} vs(\tfrac{e}{v}, \tfrac{n}{v}) + e - \mu n + pv \,.)$$

An example of a cyclic machine that can in principle transform heat into work and conversely is the Carnot machine. It consists of a cylinder with a gas and a moveable piston. It can be expanded or compressed either in heat contact with the heat sources at constant temperature thereby taking in or delivering heat, or in isolation thereby delivering or receiving work from a work source and lowering or raising its temperature. It runs quasistatically, and its cycle of operations is the following:

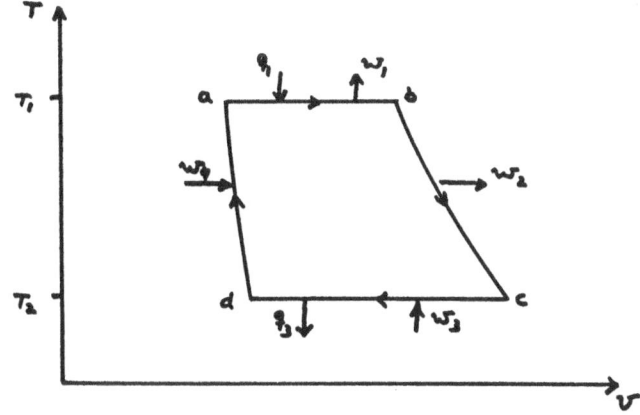

a $\rightarrow$ b: The gas is in heat contact with the hot reservoir at T_1 and with a work source, and it expands at constant temperature taking up an amount of heat q_1 and delivering an amount of work w_1 to the work source. Its entropy increases by $\beta_1 q_1$ just the amount lost by the hot source, so the process is reversible.

b → c: At b the cylinder is isolated, so no heat can be exchanged, and then it is allowed to expand reversibly in contact with the work source delivering w_2 to it. This goes on until its temperature has been lowered to T_2. Since $\delta q = 0$ all the time the entropy remains constant.

c → d: Similar to a → b, compression in heat contact with the cold reservoir at T_2. The amount of heat $q_2 = \frac{T_2}{T_1} q_1$ is delivered to it, and w_3 is taken in. Again the total entropy of the system remains constant.

d → a: Similar to b → c, compression in isolation until the temperature rises to T_1 again. w_4 is taken in, and the entropy remains constant.

The net effect of the operation of a cycle is hence that q_1 is taken in at T_1 and $w = w_1 + w_2 - w_3 - w_4$, and q_2 are delivered to the work source and the cold reservoir at T_1 respectively, in a reversible way. We see that in order to get a high efficiency one ought to try to make the operation of a heat engine as reversible as possible and the temperature ratio of the heat taken in and out as high as possible.

When e.g. a Carnot machine is run backwards it can be used as a heat pump.

E.g. we get a refrigerator if the cold reservoir is the volume we want to keep cold. and a heating device for a house e.g. if the hot reservoir is a house, and the cold one the exterior of it (at least in wintertime).

The usual approach to thermodynamics is to take as basic postulates the first law, conservation of energy and the second law, e.g. the impossibility of a cyclic perpetum mobile of the second kind. Then one can introduce the entropy and the temperature from the equation $ds = \frac{\delta q}{kT}$ for quasistatic changes, because one shows that for any cyclic such change $\int \frac{\delta q}{kT} = 0$, so the equation $s(u) - s(u_o) = \int_{u_o}^{u} \frac{\delta q}{kT}$ defines uniquely a function of the state u. Then one can go on and show that this entropy has the extremal properties we have found directly from its probabilistic definition.

We see that the probabilistic approach gives a unified derivation of the basic laws of thermodynamics from one basic assumption, the microcanonical probability law, and that in this way the mystical concept of entropy gets a natural interpretation and is related to the microscopic description of the systems.

An example of the relation between increase in entropy and loss of work:
Consider a gas in a container with a piston which is allowed to expand
from v_1 to v_2 in heat isolation:

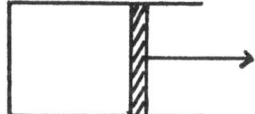

If we expand it without work the total entropy $v \cdot s(\frac{e}{v}, \frac{n}{v})$ increases
with e, n const. This increase in entropy means that we have thrown
away the possibility of getting some work, which can be obtained if
instead we let the gas expand reversibly in contact with a work source
decreasing e and having the total entropy unchanged. Let us compute
the two cases for an ideal gas. Its entropy can easily be obtained
from $g(\beta, \mu)$:

$$G_\Lambda(\beta, \mu) = \sum_0^\infty e^{\beta \mu N} \frac{v^N}{N!} \int e^{-\frac{\beta |p|^2}{2m}} \, dp \ =$$

$$= \sum_0^\infty e^{\beta \mu N} \frac{v^N}{N!} (\frac{2m\pi}{\beta})^{\frac{3N}{2}} \ =$$

$$= \exp e^{\beta \mu} \, v \, (\frac{2m\pi}{\beta})^{\frac{3}{2}} \ .$$

Hence $g(\beta, \mu)$ is $g(\beta, \mu) = e^{\beta \mu} (\frac{c}{\beta})^{\frac{3}{2}}$ $\quad (c = 2m\pi)$, and e, n are obtained
from

$$e = - g'_\beta = \frac{3}{2\beta} \cdot g \quad \text{(Derivation with } \beta\mu = \text{const.)}$$

$$n = \beta^{-1} g'_\mu = g$$

Hence $s(e, n) = g + \beta e - \beta \mu n = n + \frac{3n}{2} - n \log n \, (\frac{3n}{2ec})^{3/2} \ =$

$$= \frac{5n}{2} - \frac{5n}{2} \log n + \frac{3n}{2} \log e + \frac{3n}{2} \log \frac{2c}{3}, \quad \text{and hence}$$

$$vs(\frac{e}{v}, \frac{n}{v}) = \frac{5n}{2} - \frac{5n}{2} \log n + \frac{3n}{2} \log e + n \log v + \frac{3n}{2} \log \frac{2c}{3}.$$

We see that if v increases with e, n const. the entropy increases by

$n \log \dfrac{v_2}{v_1}$, whereas if it is constant we must have

$\dfrac{3}{2} \log e + \log v = \text{const.}$, so that

$\log \dfrac{e_1}{e_2} = \dfrac{2}{3} \log \dfrac{v_2}{v_1}$, and $w = e_1 - e_2 > 0$.

The same analysis applies if we have a dilute solution in the container and liquid outside the piston.

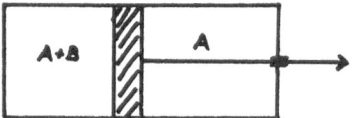

If we make a hole in the piston and pull it out without doing work the concentrations will equalize and the entropy increase, whereas if we make the piston semipermeable and let the osmotic pressure do work in a reversible expansion we can get work in the equalization process. (This could be a method of extracting work by mixing sea- and river-water reversibly.)

3.2.3. The thermodynamic description of first order phase transitions

Consider a closed system in a large container Λ described by a m. can. law defined by u and having entropy $s(u)$. Consider the state u_1 of a small but macroscopic subvolume Λ_1 (not too close to the boundary of Λ). Λ_1 also has entropy function $s(u_1)$. (Let us put $v_1 = 1$.) In theorem 5 we saw that the possible values of u_1 are those given by the can. law with $a = s'(u)$ in the limit when the size ratio of the systems goes to infinity. I.e. these values are the convex set M where the tangentplane $s = a \cdot u + g(a)$ touches Epi s:

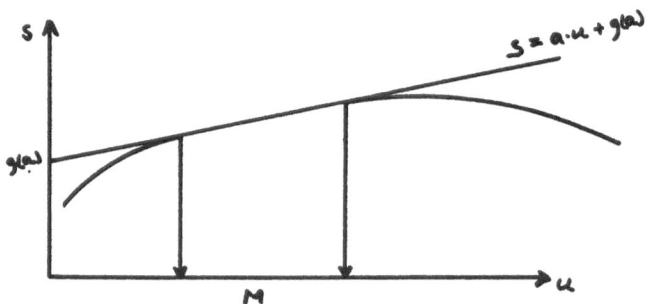

M reduces to a single point iff $g'(a)$ exists, and then $M = \{- g'(a)\}$, otherwise $s(u)$ varies linearly on M. When there is a unique u–value corresponding to a then u_1 has this value in any small subsystem Λ_1, i.e. Λ is very homogenous, and we do not find that its local properties vary much in space.

If on the other hand there is a phase transition, so Λ is filled with macroscopic regions where the local properties, density etc. are drastically different and can take a finite no. of values $u^{(1)},\ldots u^{(f)}$ characteristic of the different phases of the system (ice, water, vapour etc.), then at least if Λ_1 is chosen at random in Λ the value of u_1 should be

$$u_1 = \sum_1^f v^{(i)} u^{(i)},$$

where $v^{(i)}$ is the fraction of $|\Lambda|$ occupied by phase no. i. Because with probability $v^{(i)}$ Λ_1 will fall into the region of phase i, where u_1 takes the value $u^{(i)}$, and with negligible probability it will overlap with more than one such region. All $u^{(i)} \in M$ since $u_1 \in M$. Also

$u = \sum_1^f v^{(i)} u^{(i)}$, because $U(q)$ is essentially additive

$\frac{U(q)}{|\Lambda|} = \frac{1}{|\Lambda|} \cdot (\sum_1^f U(q^{(i)}) + \text{boundary terms})$ where $q^{(i)}$ is the part of q

located in the region of phase i, and

$$\left\langle \frac{U(q^{(i)})}{|\Lambda|} \right\rangle \approx v^{(i)} u^{(i)}.$$

If now u completely determines the probability law in Λ then also the $v^{(i)}$ are uniquely determined by u from $u = \sum_1^f v^{(i)} u^{(i)}$ when u is varied in M. This means that M ought to be a simplex in R^M, i.e. such that any point in it has a unique representation $u = \sum_1^f v^{(i)} u^{(i)}$, with

$v^{(i)} \geq 0$ $\sum_1^f v^{(i)} = 1$

(Simplies in R^2: •——•   )

If this is the case we must have $f \leq M + 1$. This is the famous Gibbs phase rule, which e.g. says that if (e,n) uniquely determine the probability law then $f \leq 3$. (There are at most three coexisting phases in an one component system, e.g. ice, water and vapour.)

If this situation occurs then it is easy to see that the average of any

other quantity $\frac{U_o}{|\Lambda|}$ will also vary linearly with $u \in M$, so that

$u_o = \sum_1^f v^{(i)} u_o^{(i)}$, as should be expected if the above picture with regions

with coexisting phases is correct.

Proof: We have shown in lemma 6 f.f. that the value of u_o is that where $\sup_{u_o} s(u_o,u) = s(u)$ is attained. If this is unique for all $u \in M$ then it is linear in $u \in M$, because if $u = \lambda u' + (1 - \lambda)u''$ e.g. and

$s(u') = s(u'_o,u')$, $s(u'') = s(u''_o,u'')$, $u_o = \lambda u'_o + (1 - \lambda)u''_o$ then because

$s(u_o,u)$ is concave $s(u_o,u) \geq \lambda s(u'_o,u') + (1 - \lambda)s(u''_o,u'') =$

$= \lambda s(u') + (1 - \lambda)s(u'') = s(u)$. (Remember that $s(u)$ is linear on M.)

Hence u_o is the maximizing value corresponding to u, q.e.d. Let us summarize these results:

If for a certain value of the intensive parameters a the set of corresponding u-values M consists of more than one point then this set ought to be a simplex if there are only finitely many phases and if u completely determines the m. can. probability law. Any point in M can be uniquely

represented as a mixture $u = \sum_1^f v^{(i)} u^{(i)}$, $v^{(i)} \geq 0$, $\sum_1^f v^{(i)} = 1$ of the

extreme points $u^{(i)}$ in M. These represent the values of u in the pure phases, and the $v^{(i)}$ are the volume fractions of the different phases.

$s(u)$ and the values of any other quantities $\frac{U_o}{|\Lambda|}$ vary linearly with u

in M, which reflects the fact that the system is filled with a mixture of pure phases in the proportions $v^{(i)}$. M consists of more than one point iff. $g'(a)$ is not defined, i.e. iff. in some direction b the directional derivatives $g'(a,b)$ are different to the right and left: $g'(a,b) \neq - g'(a,-b)$. (This is the origin of the name first order phase transition. The symptom is a jump in the first order derivatives of $g(a)$.)

It is a difficult problem to completely verify that the above picture with a mixture of different phases really occurs from first principles. This has been done with great effort by Minlos & Sinai for the Ising model.

Example 1. The Clausius-Clapeyron formula for the vapour pressure in a one component system.

Let us consider a system described by $u = (e,n)$ which has two phases
for some range of (β,μ). I.e. M is the segment $u = \lambda u^{(1)} + (1 - \lambda)u^{(2)}$
$0 \leq \lambda \leq 1$, with $u^{(i)} = (e^{(i)},n^{(i)})$ for the pure phases being functions
of (β,μ). The tangent plane $s = g(\beta,\mu) + \beta e - \beta\mu n$ is tangent to Epi s
precisely for $(e,n) \in M$, i.e. for $(e,n) \in M$ we have both

$$s(e,n) = g(\beta,\mu) + \beta e - \beta\mu n \equiv \beta(p(\beta,\mu) + e - \mu n)$$

and

$$\begin{cases} s_e'(e,n) = \beta \\ s_n'(e,n) = -\beta\mu \end{cases}$$

In particular the first equation holds for the pure phases:

$$(kT)s(e^{(i)},n^{(i)}) - e^{(i)} + \mu n^{(i)} = p(\beta,\mu)$$

If we vary (β,μ) slightly we have:

$$kdT \cdot s^{(i)} + (kT)ds^{(i)} - de^{(i)} + \mu dn^{(i)} + d\mu \cdot n^{(i)} = dp$$

but from the second equations above we have

$$ds^{(i)} = \beta(de^{(i)} - \mu dn^{(i)}) , \quad \text{or}$$

$$kTds^{(i)} = de^{(i)} - \mu dn^{(i)} , \quad \text{so}$$

$$ks^{(i)}dT + n^{(i)}d\mu = dp \quad \text{for} \quad i = 1,2 .$$

We can eliminate $d\mu$, and get

$$d\mu = \frac{ks^{(1)} - ks^{(2)}}{n^{(2)} - n^{(1)}} dT ,$$

$$dp = ks^{(1)}dT + n^{(1)} \frac{ks^{(1)} - ks^{(2)}}{n^{(2)} - n^{(1)}} dT$$

$$dp = k \left(\frac{n^{(2)}s^{(1)} - n^{(1)}s^{(2)}}{n^{(2)} - n^{(1)}} \right) dT =$$

$$dp = k \frac{\dfrac{s^{(1)}}{n^{(1)}} - \dfrac{s^{(2)}}{n^{(2)}}}{\dfrac{1}{n^{(1)}} - \dfrac{1}{n^{(2)}}} dT .$$

Here $\dfrac{s^{(i)}}{n^{(i)}}$ is the entropy per particle, and $v^{(i)} = \dfrac{1}{n^{(i)}}$ the specific

volume per particle. The latent heat per particle in the transition from one

phase to the other is $q = kT \left(\dfrac{s^{(1)}}{n^{(1)}} - \dfrac{s^{(2)}}{n^{(2)}}\right)$, so we arrive at the famous
formula

$$\frac{dp}{dT} = \frac{q}{T(v^{(1)} - v^{(2)})}$$

which expresses how the pressure changes in terms of the measurable
quantities $q, T, v^{(i)}$.

3.3. Some other uses of the concept of entropy

3.3.1. Information theory

Let us consider the basic asymptotic problem in coding theory treated
by Shannon, that of estimating how many messages have effectively to be
considered when coding the output of an information source.

For simplicity let us consider a markovian source (as Shannon did). The
messages are long sequences $x = (x_1, x_2, \ldots x_N)$ of elements from a finite
alphabet X. Let us make the m. can. markovian probability assumption
for $x \in X^N$. I.e. take as basic macrovariables $U_{ij}(x) =$ no. of consecutive
pairs i,j among $x_1, \ldots x_N$, and consider all sequences having approximatively given values of $\dfrac{U_{ij}(x)}{N}$ as equally likely: let u_{ij} be the

frequencies, and let Δ be a neighbourhood of $u = (u_{ij})$. Then consider

all sequences $x \in X^N$ having $\dfrac{U(x)}{N} \in \Delta$ equally likely with probability

$\Omega_N(\Delta)^{-1}$, $\Omega_N(\Delta) =$ the no. of such sequences. The basic interest in coding
the sequences is to estimate $\Omega_N(\Delta)$ when N is large, because only that
no. of sequences instead of $|X|^N$ need to have code words if the proba-
bility assumption is valid. Let us see that the estimation of $\Omega_N(\Delta)$
can directly be done using the methods developed before. (We can actually
think of $x \in X^N$ as the states of a one dimensional "crystal" in
$\Lambda = [1,N]$, each point $n \in \Lambda$ being occupied by an atom which can be in
one of finitely many states $x_n \in X$.)

As $N \to \infty$ we have

$$\lim_{N \to \infty} \frac{1}{N} \log \Omega_N(\Delta) = s(\Delta) = \sup_{u \in \Delta} s(u) \ ,$$

so the basic problem is to compute $s(u)$. This can be done by first considering the corresponding can. law and calculating $g(a)$. This law is defined by parameters $a = (a_{ij})$ corresponding to $U = (U_{ij})$, and

$$P_N(x|a) = G_N^{-1}(a) \ \exp - \sum_{ij} a_{ij} \ U_{ij}(x) \qquad x \in X^N \ .$$

Putting $A_{ij} = e^{-a_{ij}}$ this can be written as

$$P_N(x|a) = G_N^{-1}(a) \ A_{x_1,x_2} \ A_{x_2,x_3} \ \cdots \ A_{x_{N-1},x_N} \ , \quad \text{with}$$

$$G_N(a) = \sum_{x_1,\ldots x_N} A_{x_1,x_2} \ A_{x_2,x_3} \ \cdots \ A_{x_{N-1},x_N} = \sum_{x_1,x_N} (A^{N-1})_{x_1,x_N} \ .$$

(This is actually a markovian law, because for any $1 < n < N$ we have

$$P_N(x_{n+1}, \ \cdots \ x_N | x_1, \ldots \ x_n) = \frac{A_{x_1 x_2} \cdots A_{x_n x_{n+1}} \cdots A_{x_{N-1},x_N}}{\sum\limits_{x_{n+1} \cdots x_N} A_{x_1 x_2} \cdots A_{x_n x_{n+1}} \cdots A_{x_{N-1},x_N}} =$$

$$= \frac{A_{x_n x_{n+1}} \cdots A_{x_{N-1} x_N}}{G_{N-n}(x_n, a)} \ , \quad \text{with}$$

$$G_{N-n}(x_n, a) = \sum_{x_{n+1} \cdots x_N} A_{x_n x_{n+1}} \cdots A_{x_{N-1},x_N} = \sum_{x_N} (A^{N-n})_{x_n,x_N} \ ,$$

and we see that this conditional probability depends only on x_n, and not on $x_1, \ldots x_{n-1}$. The transition probabilities are

$$P_N(x_{n+1}|x_n) = \frac{A_{x_n,x_{n+1}} \ G_{N-n-1}(x_{n+1}, a)}{G_{N-n}(x_n, a)} \ ,$$

and the initial probabilities

$$P_N(x_1) = \frac{G_{N-1}(x_1, a)}{\sum\limits_{x_1} G_{N-1}(x_1, a)} = \frac{G_{N-1}(x_1, a)}{G_N(a)} , \quad \text{so we see that the markov chain}$$

is not stationary when N is finite).

From the expression above for $G_N(a)$ we can immediately find $g(a)$: $A = (A_{ij})$ is a matrix with positive coefficients. The Perron-Frobenius theorem then tells us that is has a largest eigenvalue $\lambda(a)$ which is positive and simple, and that the corresponding left and right eigenvectors l and r have positive components. They can be normalized so that $l \cdot r = 1$. From this follows that

$$G_N(a) = \lambda^{N-1}(a) \sum_{x_1 x_N} r_{x_1} l_{x_N} + O(\lambda_2^{N-1}) \quad \text{with} \quad \lambda_2 < \lambda(a) \quad \text{so that}$$

$$g(a) = \lim_{N \to \infty} \frac{1}{N} \log G_N(a) = \log \lambda(a)$$

From the fact that $\lambda(a)$ is simple it is not too hard to show that λ, l and r are differentiable in all A_{ij}, and that the derivatives of $\lambda(a)$ can be obtained by the formal perturbation calculation:

$$\begin{cases} Ar = \lambda r \\ (A + dA)(r + dr) = (\lambda + d\lambda)(r + dr) \end{cases} \Longrightarrow$$

$Adr + dAr \approx \lambda dr + d\lambda \cdot r$, multiply by l from the left, and remember that $l \cdot A = \lambda \cdot l$: $l \cdot dAr \approx d\lambda(l \cdot r) = d\lambda$. Hence $d\lambda = \sum_{ij} l_i \, dA_{ij} r_j$, so

$$\frac{\partial \lambda}{\partial A_{ij}} = l_i r_j \ .$$

From this follows that $g(a)$ is differentiable, and $dg = d \log \lambda =$

$$= \lambda^{-1} d\lambda = \lambda^{-1} \sum_{ij} l_i \, dA_{ij} \, r_j = -\lambda^{-1} \sum_{ij} l_i A_{ij} \, da_{ij} \, r_j \ , \quad \text{so that}$$

$$\frac{\partial g}{\partial a_{ij}} = -\lambda^{-1} l_i A_{ij} r_j \ .$$

The correspondance between u and a which says that $u = -g'(a)$ if the can. law is centered at u hence gives

$$\lambda^{-1} l_i A_{ij} r_j = u_{ij} \ .$$

Let us introduce P_{ij} and p_i by $P_{ij} = \lambda^{-1} r_i^{-1} A_{ij} r_j$, and $p_i = l_i r_i$. We then have $P_{ij} > 0$, $\sum_j P_{ij} = 1$, $\sum_i p_i = 1$, $\sum_i p_i P_{ij} = p_j$, so

these quantities define a stationary markov chain with the right pair frequencies:

$$u_{ij} = p_i \cdot P_{ij} \; .$$

(Actually, the non stationary transition probabilities obtained above for finite N converge to P_{ij} as $N \to \infty$. This is easy to see, because

$$G_{N-n}(x_n, a) = \lambda^{N-n} \; r_{x_n} \cdot \sum_{x_N} 1_{x_N} + O(\lambda_2^{N-n}) \; , \quad \text{so that}$$

$$\frac{G_{N-n-1}(x_{n+1}, a)}{G_{N-n}(x_n, a)} \to \lambda^{-1} \; r_{x_n}^{-1} \; r_{x_{n+1}} \; , \quad \text{and}$$

$$P_N(x_{n+1} | x_n) \to \lambda^{-1} \; r_{x_n}^{-1} \; A_{x_n, x_{n+1}} \; r_{x_{n+1}} = P_{x_n x_{n+1}} \quad \text{as} \quad N \to \infty \; .$$

For the initial probabilities $P_N(x_1)$ similarly we have

$$P_N(x_1) \to \frac{r_{x_1}}{\displaystyle\sum_{x_1} r_{x_1}} \ne p_{x_1} \; , \quad \text{but} \quad P_N(x_n) \to p_{x_n}$$

when first $N \to \infty$ and then $n \to \infty$.)

We can now calculate $s(u)$ from

$$s(u) = \inf_a \; (g(a) + a \cdot u) = g(a) + a \cdot u \quad \text{when} \quad u = - g'(a) \; . \quad \text{I.e.}$$

$$s(u) = \log \lambda - \sum_{ij} u_{ij} \log A_{ij} = - \sum_{ij} u_{ij} \log \frac{A_{ij}}{\lambda} =$$

$$= - \sum_{ij} p_i P_{ij} \log r_i P_{ij} r_j^{-1} = - \sum_{ij} p_i P_{ij} \log P_{ij} +$$

$$+ \sum_{ij} p_i P_{ij} (\log r_i - \log r_j) \; .$$

The last term is $\displaystyle\sum_i p_i \log r_i - \sum_j p_j \log r_j = 0$, so we see that

$$s(u) = h(P) = - \sum_{ij} p_i P_{ij} \log P_{ij} \; , \quad \text{which is Shannons famous formula.}$$

Let us sum up our basic results: We have considered the m. can. and can. laws defined by $U_{ij}(x)$. For finite N the latter is a non stationary markov chain whose probabilities converge to those of the stationary markov chain defined above by p_i, P_{ij} having $p_i P_{ij} = u_{ij}$.

Since $g(a)$ is differentiable our basic result about equivalence of ensembles, Theorem 3, tels us that for any neighbourhood $\Delta \ni u$ we have

$$P_N \left(\frac{U(x)}{N} \in \Delta \mid a\right) \to 1 \quad \text{if} \quad u = - g'(a) ,$$

and the number of such sequences $\Omega_N(\Delta)$ increases as $e^{N \cdot s(u)} = e^{N \cdot h(P)}$

in the sense that $\frac{1}{N} \log \Omega_N(\Delta) = \sup_{u \in \Delta} s(u)$, so

$$\lim_{\Delta \downarrow u} \lim_{N \to \infty} \frac{1}{N} \log \Omega_N(\Delta) = s(u) = h(P) .$$

This is Mc. Millan's theorem for this particular markovian case.

If we consider the interpretation of the above system as a one dimensional "crystal" the "energy" $\sum_{ij} a_{ij} U_{ij}(x) = a_{x_1 x_2} + a_{x_2 x_3} + \ldots + a_{x_{N-1} x_N}$ is a sum of contributions from nearest neighbours. The basic result that $g(a)$ is differentiable then means that such a system does not exhibit any phase transition. This result is easily generalized to any finite range inter-action using similar arguments as above.

3.3.2. Statistical models

As an example let us consider the so called Wilson model for predicting traffic flows between a given set of origins and destinations in an urban area. In this field it is very fashionable to use entropy arguments to derive several formulas.

Consider the daily traffic from a given set of origins $i = 1, \ldots n$ to a given set of destinations $j = 1, \ldots m$. There are N persons each making such a trip, and one is interested in predicting the total flows U_{ij} between all o-d pairs. A microscopic state description is a list $x = \{(i_n, j_n) \quad n = 1, \ldots N\}$ of the o-d pairs of each individual person. For a given set of U_{ij} with $\sum_{ij} U_{ij} = N$ the total no. of microstates having these flows is

$$\Omega_N(u) = \frac{N!}{\prod_{ij} U_{ij}!} \qquad u_{ij} = \frac{U_{ij}}{N}$$

By Stirling's formula, $N! \approx (\frac{N}{e})^N$, this means that

$$s(u) = \lim_{N \to \infty} \frac{1}{N} \log \Omega_N(u) = - \sum_{ij} u_{ij} \log u_{ij}$$

as $N \to \infty$ $u_{ij} = \dfrac{U_{ij}}{N} = \text{const.}$

One common m. can. probability assumption is to assume that all x having given values of $O_i(x) = \sum_j U_{ij}(x)$ and $D_j(x) = \sum_i U_{ij}(x)$ and

of $T(x) = \sum_{ij} l_{ij} U_{ij}(x) =$ the total travelled distance are equally likely. The corresponding $\Omega_N(o,d,t)$ and entropy $s(o,d,t)$ is then then obtained from $\Omega_N(o,d,t) = \sum_{\substack{\sum_j u_{ij} = o_i \\ \sum_i u_{ij} = d_j \\ \sum_{ij} l_{ij} u_{ij} = t}} \Omega_N(u)$,

and the principle that "only the largest term contributes" gives

$$s(\boldsymbol{o},d,t) = \sup_u - \sum_{ij} u_{ij} \log u_{ij} \quad \text{when}$$

$$\sum_j u_{ij} = o_i$$

$$\sum_i u_{ij} = d_j$$

$$\sum_{ij} l_{ij} u_{ij} = t$$

$$u_{ij} \geq 0$$

The argument used extensively before then says that the probability density of $\dfrac{U_{ij}(x)}{N}$ in this ensemble is $\dfrac{\Omega_N(u)}{\Omega_N(o,d,t)}$, so that in the

limit $N \to \infty$ all probability mass of $\dfrac{U(x)}{N}$ will be concentrated to the values giving the $\sup_u s(u)$ above. Hence this maximization principle gives the values of u_{ij} predicted by the m. can. law in the limit $N \to \infty$. The corresponding can. law is defined by parameters a_i, b_j, c corresponding to O_i, D_j, T and has density

$$P_N(x|a,b,c) = G_N^{-1} \exp - \sum_{ij} (a_i + b_j + cl_{ij}) U_{ij}(x)$$

$$G_N(a,b,c) = \sum_x \exp - \sum_{ij} (a_i + b_j + cl_{ij}) U_{ij}(x) =$$

$$= \sum_U \frac{N!}{\prod_{ij} U_{ij}!} \prod_{ij} e^{-(a_i+b_j+cl_{ij})U_{ij}} = (\sum_{ij} e^{-(a_i+b_j+cl_{ij})})^N .$$

The can. density of U_{ij} is then multinomial:

$$P_N(U|a,b,c) = \frac{N!}{\prod_{ij} U_{ij}!} \prod_{ij} P_{ij}^{U_{ij}}$$

with $P_{ij} = \dfrac{e^{-(a_i+b_j+cl_{ij})}}{\sum_{ij} e^{-(a_i+b_j+cl_{ij})}}$.

We see that $g(a,b,c) = \log (\sum_{ij} e^{-(a_i+b_j+cl_{ij})})$ which is differentiable,

so our Theorem 3 tells us that if the can. law is centered correctly,
i.e. if

$$- g'_{a_i} = o_i$$

$$- g'_{b_j} = d_j$$

$$- g'_{c} = t$$

then the averages of $\dfrac{U_{ij}}{N}$ which are $P_{ij}(a,b,c)$ will be the same as the
m. can. averages defined by o,d,t. The centering equations are:

$$\sum_j e^{-(a_i+b_j+cl_{ij}+g)} = o_i$$

$$\sum_i e^{-(a_i+b_j+cl_{ij}+g)} = d_j$$

$$\sum_{ij} l_{ij} e^{-(a_i+b_j+cl_{ij}+g)} = t$$

These non linear equations have to be solved by some iterative procedure
to give a,b,c and then $P_{ij}(a,b,c)$.

Reference: A.G. Wilson. Entropy in Urban and Regional Modelling.
Pion, London (1970).

3.4. Proof of the fundamental asymptotic properties of the structure measure in the thermodynamic limit

Let us now return to the proof of Theorem 2, which establishes the basic asymptotic property of the structure measure:

$$\lim_{\Lambda \to \infty} \frac{1}{|\Lambda|} \log \Omega_\Lambda(A,a) = s(A,a) = \sup_{u \in A} (s(u) - a \cdot u) \,, \quad A \text{ being an open}$$

convex subset of R^M.

We first recall the basic definitions and assumptions in section 3.1. preceding the formulation of Theorem 2. The above limit is usually estab-lished by an argument based on superadditivity as follows. For any bounded region Λ let $\Lambda' \subset \Lambda$ be the set of points in Λ with distance at least R to Λ^c, and define $\Omega'_\Lambda(A,a)$ as $\Omega_\Lambda(A,a)$ but with the extra restric-tion that all particles are restricted to be in Λ':

$$\Omega'_\Lambda (A,a) = \int_{\substack{\frac{U(q)}{|\Lambda|} \in A \\ q \in \Gamma_{\Lambda'}}} e^{-a \cdot U(q)} \, \omega(dq) = \Omega_{\Lambda'}\left(\frac{|\Lambda|}{|\Lambda'|} A, a\right) \,.$$

$(\Omega'_\Lambda = 0$ if Λ' is empty$)$.

Then $\log \Omega'_\Lambda (A,a)$ is superadditive in Λ:

If $\Lambda = \Lambda_1 \cup \Lambda_2$ with Λ_1, Λ_2 disjoint then $\Omega'_\Lambda(A,a) \geq \Omega'_{\Lambda_1}(A,a) \cdot \Omega'_{\Lambda_2}(A,a)$, because if in Ω'_Λ we only consider configurations where all particles are in $\Lambda'_1 \cup \Lambda'_2$, so $q = (q_1, q_2)$ with $q_i \in \Lambda'_i$ then $U(q) = U(q_1) + U(q_2)$ because q_1 and q_2 are at least a distance R apart and

$$\frac{U(q)}{|\Lambda|} = \frac{|\Lambda_1|}{|\Lambda|} \frac{U(q_1)}{|\Lambda_1|} + \frac{|\Lambda_2|}{|\Lambda|} \frac{U(q_2)}{|\Lambda_2|} \in A \quad \text{if} \quad \frac{U(q_i)}{|\Lambda_i|} \in A \,,$$

so just as in the proof of d) in Lemma 4

$$\Omega'_\Lambda (A,a) \geq$$

$$\geq \sum_N \frac{1}{N!} \sum_{N_1+N_2=N} \binom{N}{N_1} \int_{\substack{\frac{U(q_1)}{|\Lambda_1|} \in A \\ q_1 \in (\Lambda'_1)^{N_1}}} e^{-a \cdot U(q_1)} \, dq_1 \int_{\substack{\frac{U(q_2)}{|\Lambda_2|} \in A \\ q_2 \in (\Lambda'_2)^{N_2}}} e^{-a \cdot U(q_2)} \, dq_2 =$$

$$= \Omega'_{\Lambda_1} (A,a) \cdot \Omega'_{\Lambda_2} (A,a).$$

First we let Λ be a cube Λ_ℓ with side $L \cdot 2^\ell$ and volume
$|\Lambda_\ell| = |\Lambda_0| 2^{\ell d}$, $\ell = 0, 1, \ldots$, Then from the superadditivity follows that

$s_\ell = \dfrac{1}{|\Lambda_\ell|} \log \Omega'_{\Lambda_\ell}(A, a)$ increases with ℓ, because $\Lambda_{\ell+1}$ is the union

of 2^d translates of Λ_ℓ, so $\Omega'_{\Lambda_{\ell+1}}(A, a) \geq (\Omega'_{\Lambda_\ell}(A, a))^{2^d}$ and hence

$s_{\ell+1} \geq s_\ell$. The limit of s_ℓ thus exists, call it

$$s(A, a) = \lim_{\ell \to \infty} s_\ell = \sup_{\ell > 0} s_\ell .$$

$s(A, a)$ will have the regularity required: $s(A, a) = \sup\limits_{C \subset A} s(C, a)$, where C

is open convex with compact closure $C \subset A$. This is easily seen first if
$s(A, a) = -\infty$. If $s(A, a) > -\infty$, for any $\varepsilon > 0$ take Λ_ℓ such that

$$\frac{1}{|\Lambda_\ell|} \log \Omega'_{\Lambda_\ell}(A, a) > s(A, a) - \frac{\varepsilon}{2} .$$

$\Omega'_{\Lambda_\ell}(A, a)$ being a Borel measure on R^M has the above regularity, so there
is a C such that

$$\frac{1}{|\Lambda_\ell|} \log \Omega'_{\Lambda_\ell}(C, a) > \frac{1}{|\Lambda_\ell|} \log \Omega'_{\Lambda_\ell}(A, a) - \frac{\varepsilon}{2} ,$$

and hence

$$s(C, a) \geq \frac{1}{|\Lambda_\ell|} \log \Omega'_{\Lambda_\ell}(C, a) > s(A, a) - \varepsilon .$$

We can now directly prove that the properties described in Lemma 4 a)-d)
are valid for $s(A, a)$: a) and b) are clear. To prove c) suppose that
$A = A_1 \cup A_2$ with $s(A_1, a) \geq s(A_2, a)$ e.g. Then

$$\Omega'_{\Lambda_\ell}(A_1, a) \leq \Omega'_{\Lambda_\ell}(A, a) \leq \Omega'_{\Lambda_\ell}(A_1, a) + \Omega'_{\Lambda_\ell}(A_2, a) \leq 2 \exp |\Lambda_\ell| s(A_1, a) ,$$

so if $s(A_1, a) = -\infty$ then all terms are zero and $s(A, a) = -\infty$. If
$s(A_1, a) > -\infty$ then

$$\frac{1}{|\Lambda_\ell|} \log \Omega_{\Lambda_\ell}(A_1, a) \to s(A_1, a)$$

and hence

$$\frac{1}{|\Lambda_\ell|} \log \Omega_{\Lambda_\ell}(A, a) \to s(A_1, a) \text{ too, i.e. } s(A, a) = s(A_1, a) .$$

To prove d), for any $\varepsilon > 0$ take Λ_ℓ such that $\frac{1}{|\Lambda_\ell|} \log \Omega'_{\Lambda_\ell} (A_i,a) >$

$> s(A_i,a) - \varepsilon$ for $i = 1,2$. (The larger of ℓ_1 and ℓ_2 needed for A_1 and A_2 will do.) Decompose $\Lambda_{\ell+1}$ into 2^d translates of Λ_ℓ, and call them $\Lambda^{(1)},\ldots\Lambda^{(2^d)}$. Consider only configurations in $\Lambda_{\ell+1}$ of the form $q = (q^{(1)},\ldots q^{(2^d)})$ with $q^{(i)} \in \Gamma_{\Lambda^{(i)}}$, and with $\frac{U(q^{(i)})}{|\Lambda_\ell|} \in A_1$ or A_2 respectively for half of the boxes. Then

$$\frac{U(q)}{|\Lambda_{\ell+1}|} = \frac{|\Lambda_\ell|}{|\Lambda_{\ell+1}|} \left(\sum_1^{2^{d-1}} \frac{U(q^{(i)})}{|\Lambda_\ell|} + \sum_{2^{d-1}+1}^{2^d} \frac{U(q^{(i)})}{|\Lambda_\ell|} \right) \in \frac{A_1 + A_2}{2} ,$$

so we get a lower bound to $\Omega'_{\Lambda_{\ell+1}} (\frac{A_1 + A_2}{2}, a)$ by considering only such

configurations. Hence

$$s(\frac{A_1 + A_2}{2}, a) \geq \frac{1}{|\Lambda_{\ell+1}|} \log \Omega'_{\Lambda_{\ell+1}} (\frac{A_1 + A_2}{2}, a) \geq$$

$$\geq \frac{1}{2|\Lambda_\ell|} \log \Omega'_{\Lambda_\ell} (A_1,a) + \frac{1}{2|\Lambda_\ell|} \log \Omega'_{\Lambda_\ell} (A_2,a) \geq$$

$$\geq \frac{1}{2} s(A_1,a) + \frac{1}{2} s(A_2,a) - \varepsilon, \quad \text{with} \quad \varepsilon \quad \text{arbitrary.}$$

(If $s(A_1,a)$ or $s(A_2,a) = -\infty$ d) is trivially true.)

On the basis of Lemma 4 we see that Lemma 5 is valid for $s(A,a)$, i.e.

$$s(A,a) = \sup_{u \in A} (s(u) - a \cdot u) ,$$

$s(u)$ having the properties described in Lemma 5.

In order to establish the existence of the limit for more general regions $\Lambda \to \infty$ we have to introduce a condition on Λ saying that "boundary effects can be neglected" as $\Lambda \to \infty$.

Definition: A sequence $\{\Lambda_i\}$ is said to tend to infinity in the sense of van Hove if $|\Lambda_i| \to \infty$, and if $\frac{|\Lambda_i(d)|}{|\Lambda_i|} \to 0$ for any d, where $\Lambda_i(d)$ is the set of points with distance at most d from the boundary of Λ_i.

Lemma 11. If $\Lambda \to \infty$ in the sense of van Hove then

$$\lim_{\Lambda \to \infty} \frac{1}{|\Lambda|} \log \Omega_\Lambda(A,a) \geq \lim_{\Lambda \to \infty} \frac{1}{|\Lambda|} \log \Omega'_\Lambda(A,a) \geq s(A,a) .$$

Proof: We try to approximate $\Omega'_\Lambda(A,a)$ from below by something which converges to $s(A,a)$. To this end fill Λ' with many translates of Λ_ℓ. E.g. consider R^d as the union of all such translates with the vertices being integer multiples of $L \cdot 2^\ell$. Those which cover the boundary of Λ' are contained in $\Lambda(D+R)$ if D is larger than the diameter of Λ_ℓ. Let N_ℓ be the number of those which are inside Λ', and let Λ_0 be their union, $\Lambda_0 = \bigcup_1^{N_\ell} \Lambda_i$. Then $|\Lambda'| - |\Lambda_0| \leq$ the volume of those covering $\partial\Lambda' \leq |\Lambda(D+R)|$, and $|\Lambda| - |\Lambda'| \leq |\Lambda(R)|$, so $1 - \frac{|\Lambda_0|}{|\Lambda|} \to 0$, and $\frac{N_\ell |\Lambda_\ell|}{|\Lambda|} \to 1$.

Take any $C \subset A$ as in the definition of regularity above. Then for configurations of particles $q_i \in \Gamma_{\Lambda_i'}$ the combined configuration q is $\in \Gamma_{\Lambda'}$, and if $\frac{U(q_i)}{|\Lambda_\ell|} \in C$ then

$$\frac{U(q)}{|\Lambda|} = \frac{1}{|\Lambda|} \sum_i U(q_i) \in \sum_i \frac{|\Lambda_\ell|}{|\Lambda|} C = \frac{N_\ell |\Lambda_\ell|}{|\Lambda|} C \subset A$$

if Λ is big enough, so that $\frac{N_\ell |\Lambda_\ell|}{|\Lambda|}$ is close to 1.

Hence integrating only over such configurations $q \in \Gamma_\Lambda$ we get the lower bound $\Omega'_\Lambda(A,a) \geq (\Omega'_{\Lambda_\ell}(C,a))^{N_\ell}$ when Λ is big enough. This gives

$$\frac{1}{|\Lambda|} \log \Omega'_\Lambda(A,a) \geq \frac{1}{|\Lambda_\ell|} (\log \Omega'_{\Lambda_\ell}(C,a)) \frac{N_\ell |\Lambda_\ell|}{|\Lambda|}$$

and as $\Lambda \to \infty$:

$$\lim_{\Lambda \to \infty} \geq \frac{1}{|\Lambda_\ell|} \log \Omega'_{\Lambda_\ell}(C,a) \qquad \text{for all } \ell .$$

Hence

$$\varliminf_{\Lambda \to \infty} \geq s(C,a) \qquad \text{for all } C \subset A \;,$$

and finally

$$\varliminf_{\Lambda \to \infty} \geq s(A,a) = \sup_{C \subset A} s(C,a)$$

In order to get an estimate from above we make another regularity assumption about Λ, which is somewhat restrictive but adequate for most common shapes.

Definition: $\{\Lambda_i\}$ tending to infinity in the sense of van Hove is said to be approximable by cubes if one can find such $\bar{\Lambda}_i \supset \Lambda_i$ with

$$\lim_{i \to \infty} \frac{|\Lambda_i|}{|\bar{\Lambda}_i|} = \inf_i d{:}\varrho = c > 0 \; . \quad \text{(E.g. if } \Lambda_i \text{ is a sphere of radius } i$$

then the condition is true, but if Λ_i is a spherical shell of radius i and thickness $\sqrt{i}$ it is not.)

The side of $\bar{\Lambda}_i$, L_i, can always be taken of the form $\bar{L}_i = L{\cdot}2^{\ell_i}$,

because we can always take $\ell_i = \left[\log_2\left(\dfrac{L_i}{L}\right)\right] + 1$ if necessary, and

$\dfrac{\cdot L_i}{\bar{L}_i} \geq 2^{-2} = \dfrac{1}{4}$, decreasing c to at most $c{\cdot}4^{-d}$. By enlarging if necessary

$\bar{\Lambda}_i$ we can always take $\dfrac{|\Lambda_i|}{|\bar{\Lambda}_i|} \leq \dfrac{1}{2}$ e.g. and then $\bar{\Lambda}_i \diagdown \Lambda_i \to \infty$ in the sense

of van Hove.

Lemma 12. If $\Lambda \to \infty$ in the sense of van Hove and is approximable by cubes then

$$\varlimsup_{\Lambda \to \infty} \frac{1}{|\Lambda|} \log \Omega_\Lambda(A,a) \leq s(A,a) \quad \text{if} \quad s(A,a) > -\infty$$

or if $s(A,a) = -\infty$ and A has positive distance to the domain of $s(u)$ $D = \{u; \; s(u) > -\infty\}$. In the latter case $\Omega_\Lambda(A,a) = 0$ for all Λ .

Proof: Consider first the case $s(A,a) > -\infty$, and take cubes $\Lambda_1' \supset \Lambda$ as in the definition above such that $\Lambda_1' \supset \Lambda$, and put $\Lambda_2 = \Lambda_1 \diagdown \Lambda$.

If we consider only configurations having particles only in $\Lambda \cup \Lambda_2'$ we get a lower bound for Ω_{Λ_1}' (A,a):

$$\Omega_{\Lambda_1}' \ (A,a) \geq \Omega_{\Lambda_2}' \ (A,a) \cdot \Omega_{\Lambda}(A,a) \ , \quad \text{and}$$

$$\frac{1}{|\Lambda_1|} \log \Omega_{\Lambda_1}' \ (A,a) \geq \frac{|\Lambda|}{|\Lambda_1|} \frac{1}{|\Lambda|} \log \Omega_{\Lambda}(A,a) + \frac{|\Lambda_2|}{|\Lambda_1|} \frac{1}{|\Lambda_2|} \log \Omega_{\Lambda_2}' \ (A,a) \ .$$

As $\Lambda \to \infty$ the left side tends to $s(A,a)$, so using Lemma 11 for Ω_{Λ_2}' we get

$$s(A,a) \geq c \ \overline{\lim_{\Lambda \to \infty}} \ \frac{1}{|\Lambda|} \log \Omega_{\Lambda}(A,a) + (1 - c)s(A,a) \ , \quad \text{and since}$$

$$c > 0 \quad s(A,a) \geq \overline{\lim_{\Lambda \to \infty}} \ .$$

If $s(A,a) = -\infty$ then $\Omega_{\Lambda_1}' \ (A,a) = 0$, but we can not conclude that $\Omega_{\Lambda}(A,a) = 0$ unless we know that $\Omega_{\Lambda_2}' \ (A,a) > 0$. We therefore modify the argument by considering configurations (q,q_2) in $\Lambda \cup \Lambda_2'$ with $\frac{U(q)}{|\Lambda|} \in A$, $\frac{U(q_2)}{|\Lambda_2|} \in A_2$, where A_2 is a small neighbourhood of a point u in D, so we know that $s(A_2,a) > -\infty$. Then we know that for the total U we have

$$\frac{U(q) + U(q_2)}{|\Lambda_1|} \in \frac{|\Lambda|}{|\Lambda_1|} A + \frac{|\Lambda_2|}{|\Lambda_1|} A_2 \subseteq A_c \ ,$$

where the convex set A_c is defined as $A_c = \{\lambda u + (1-\lambda)u_2; \text{ with } u \in A, u_2 \in A_2, \lambda \geq c\}$. Hence the previous bound is replaced by $\Omega_{\Lambda_1}' \ (A_c,a) \geq \Omega_{\Lambda_2}' \ (A_2,a) \cdot \Omega_{\Lambda}(A,a).$

Lemma 11 implies that $\Omega_{\Lambda_2}' \ (A_2,a) > 0$ if Λ_2 is sufficiently large since $\Lambda_2 \to \infty$ in the sense of van Hove and $s(A_2,a) > -\infty$. Hence if we can choose A_2 such that $s(A_c,a) = -\infty$ also we see that $\Omega_{\Lambda_1}' \ (A_c,a) = 0$ and hence $\Omega_{\Lambda}(A,a) = 0$, and

$$\overline{\lim_{\Lambda \to \infty}} \ \frac{1}{|\Lambda|} \log \Omega_{\Lambda}(A,a) = -\infty.$$

To find a good choice of A_2 let $d = d(A,D) > 0$, and take $u \in D$ such that $d(u,A) \leq d + \frac{cd}{3}$ e.g. and let A_2 be a sphere of radius $\frac{cd}{3}$ around u. Then we claim that $d(A_c,D) \geq \frac{cd}{3} > 0$, so A_c is disjoint from D and hence $s(A_c,a) = \sup_{u \in A_c} \ (s(u) - a \cdot u) = -\infty$.

In fact if $v_\lambda = \lambda v + (1-\lambda)v_2$ with $v \in A$, $v_2 \in A_2$, $\lambda \geq c$, and $u_\lambda = \lambda v + (1-\lambda)u$ then $|v_\lambda - u_\lambda| \leq (1-\lambda)|v_2 - u| \leq \frac{cd}{3}$, and for any $w \in D$ we have

$$|w - u_\lambda| + |u_\lambda - v| \geq |w - v| \geq d \quad \text{and}$$

$$|w - v_\lambda| + |v_\lambda - u_\lambda| \geq |w - u_\lambda|.$$

Since $|u_\lambda - v| = (1-\lambda)(u-v) \leq (1-c)(d + \frac{cd}{3}) \leq (1-c)d + \frac{cd}{3}$ we see that

$$|w - v_\lambda| \geq d - |u_\lambda - v| - |v_\lambda - u_\lambda| \geq d - (1-c)d - \frac{cd}{3} \frac{cd}{3} = \frac{cd}{3}$$

as claimed above.

Lemma 11 and 12 together establish

__Theorem 2.__If Λ tends to infinity in the sense of van Hove and is approximable by cubes then $\lim\limits_{\Lambda \to \infty} \frac{1}{|\Lambda|} \log \Omega_\Lambda(A,a) = s(A,a)$ exists if $s(A,a) > - \infty$, or if $s(A,a) = - \infty$ and $d(A,D) > 0$. ($s(A,a) = + \infty$ possibly.)

$s(A,a)$ is inner regular:

$s(A,a) = \sup\limits_{C \subset A} s(C,a)$, where C is open convex with compact closure $\subset A$,

and it has the properties described in Lemma 4. If $s(A,a) = - \infty$ and $d(A,D) > 0$ then $\Omega_\Lambda(A,a) = 0$ for Λ sufficiently large.

(The fact that we have to make the extra assumption $d(A,D) > 0$ when $s(A) = - \infty$ does not influence the use of $s(u)$ in the discussion following Lemma 6.)

At this stage we can prove the characterization of D in Lemma 5 d):

$$\bar{D} = \bar{R} , \quad \text{where}$$

$$R = \bigcup_\Lambda \{\text{the support of } \Omega_\Lambda(du)\}$$

__Proof:__ If $u \notin \bar{R}$ then the same is true for some open $A \ni u$, so that $\Omega_\Lambda(A) = 0$ for all Λ , and by Lemma 11 $s(A) = - \infty$, so $A \subset D^c$ and $u \notin \bar{D}$. Hence $\bar{D} \subseteq \bar{R}$.
If $u \in R$ then for some Λ u is in the support of $\Omega_\Lambda(du)$, i.e. $\Omega_\Lambda(A) > 0$ for all open $A \ni u$. From Lemma 12 then follows that for all such A either $s(A) > - \infty$ or $d(A,D) = 0$. This implies that $u \in \bar{D}$, because otherwise for some open $A \ni u$ $d(A,D) = d(A,\bar{D}) > 0$, and $s(A) = - \infty$. Hence $R \subseteq \bar{D}$, and therefore $\bar{R} \subseteq \bar{D}$.

3.4.1. Properties of the entropy s(e,n)

We now consider the most important special case when $U = (H,N)$ and $a = (\beta,-\beta\mu)$ and we will give conditions to ensure that

$\quad g(\beta,\mu) = \sup_{e,n} (s(e,n) - \beta e + \beta\mu n)$ is finite.

We will also prove that $s(e,n)$ is increasing in e and differentiable in (e,n). Since $\beta = s'_e(e,n)$ as remarked in 3.2.1 $s(e,n)$ being increasing in e is equivalent to β being positive, which is needed to have a "physical" behaviour of the system.

Consider first the situation when $U_1(q) = $ total potential energy and $U_2(q) = N$, and call the correponding functions $\bar{s}(e,u)$, $\bar{g}(\beta,\mu)$. The natural condition to ensure that $\bar{g}(\beta,\mu)$ is finite for $\beta \geq 0$, μ arbitrary is the following:

Definition: $U_1(q)$ is called stable if $U_1(q) \geq - K \cdot N$ for all $q \in R^{Nd}$ and some $K \geq 0$. Stability implies that $\bar{g}(\beta,\mu)$ is finite, because we have the bound:

$$\bar{G}_\Lambda(\beta,\mu) = \sum_{N \geq 0} \int_{q \in \Lambda^N} e^{-\beta U_1(q)+\beta\mu N} \frac{dq}{N!} \leq$$

$$\leq \sum_{N \geq 0} e^{(\beta K+\beta\mu)N} \frac{|\Lambda|^N}{N!} = \exp |\Lambda| e^{\beta K+\beta\mu} \quad,$$

so that

$$\bar{g}(\beta,\mu) = \lim_\Lambda \frac{1}{|\Lambda|} \log \bar{G}_\Lambda(\beta,\mu) \leq e^{\beta K+\beta\mu} \quad.$$

We also have $\bar{g}(\beta,\mu) \geq 0$ because $G_\Lambda(\beta,\mu) \geq 1$.

(In Ruelle: Statistical Mechanics criteria are given for a pair interaction $U_1(q) = \sum_{i<j} u(q_i-q_j)$ to be stable, and it is shown that if $u(q)$

is u.s.c. then stability is also necessary in order to have $\bar{G}_\Lambda(\beta,\mu)$ finite for any bounded Λ.) We now want to include also the kinetic energy $U_0(p) = \sum_1^N \dfrac{|p_i|^2}{2m}$ in the energy and replace $U_1(q)$ by $H(p,q) = U_0(p) + U_1(q)$. In order to do this include also the momenta $p = (p_1, \ldots p_N) \in R^{Nd}$ among the coordinates, so the state is described by $(p,q) \in \bigcup_{N>0} R^{Nd} \times \Lambda^N$ with the basic measure $dp\omega_N(dq)$, and add $U_0(p)$ to the other observables and define the extended structure measure for $A \subset R^3$:

$$\Omega_\Lambda(A,a) = \sum_{\substack{N \geq 0 \\ |\Lambda|^{-1}(U_0,U_1,U_2) \in A \\ (p,q) \in R^{Nd} \times \Lambda^N}} \int e^{-\beta U_0(p) - \beta U_1(q) + \beta\mu N} \, dp\omega_N(dq) \; .$$

$(a = (\beta, -\beta\mu)$, $U_2 = N.)$

Also, define $\Omega'_\Lambda(A,a)$ as before with the restriction that $q \in \Gamma_{\Lambda'}$. Then $\log \Omega'_\Lambda(A,a)$ will be superadditive just as before, because there is no coupling between different p_i or between p_i and q_j, so $U(p,q)$ is still additive for configurations (p,q) having the q's separated by at least R. All the arguments of 3.4. showing that $\lim_{\Lambda \to \infty} \dfrac{1}{|\Lambda|} \log \Omega_\Lambda(A,a) = s(A,a)$, with $s(A,a)$ having the properties described by Lemma 4 and 5 can then be applied to the extended $\Omega_\Lambda(A,a)$, except that the bound of Lemma 4, a) is not valid, because the p_i vary over all of R^d. Instead we get following bound: Suppose that $u_0 \leq c$ when $(u_0, u_1, u_2) \in A$, then

$$\Omega_\Lambda(A,a) \leq \sum_{N \geq 0} \int_{U_0(p) \leq c|\Lambda|} e^{-\beta U_0(p)} \, dp \int_{q \in \Lambda^N} e^{(\beta K + \beta\mu)N} \, \frac{dq}{N!} \leq$$

$$\leq \sum_{\substack{N \geq 0 \\ |p|^2 \leq 2mc|\Lambda| \\ p \in R^{Nd}}} \int dp \; e^{(\beta K + \beta\mu)N} \, \frac{|\Lambda|^N}{N!} =$$

$$= \sum_{N \geq 0} \frac{(2\pi mc|\Lambda|)^{\frac{Nd}{2}}}{(\frac{Nd}{2})!} \, \frac{|\Lambda|^N}{N!} \, e^{(\beta K + \beta\mu)N} \leq \sup_{N \geq 0} \frac{(2\pi mc|\Lambda|)^{\frac{Nd}{2}}}{(\frac{Nd}{2})!} \cdot \exp|\Lambda| \, e^{\beta K + \beta\mu} \; .$$

But $x! \geq (\frac{x}{e})^x$ for $x \geq 0$, so

$$\sup_{N \geq 0} \leq \sup_{x > 0} (\frac{eb}{x})^x = e^b \quad \text{for} \quad b = 2\pi mc|\Lambda|, \quad \text{and}$$

$$\Omega_\Lambda(A,a) \leq \exp|\Lambda|(2\pi mc + e^{\beta K + \beta \mu}) .$$

Hence the bound of Lemma 4 a) can be replaced by

$$s(A,a) \leq (2\pi m) \sup_{u_0 \in A} u_0 + e^{\beta K + \beta \mu}, \quad \text{and}$$

$$s(u) \leq (2\pi m)u_0 + 1 ,$$

so $s(u) < + \infty$ always.

$s(u_0,u_1,u_2)$ can be expressed in terms of the "potential entropy" $\bar{s}(u_1,u_2)$ as follows:

Let $A = \Delta_0 \times \Delta_1 \times \Delta_2$ be a small neighbourhood of (u_0,u_1,u_2). Then

$$\Omega_\Lambda(A) = \sum_{\frac{N}{|\Lambda|} \in \Delta_2} \int_{\frac{|p|^2}{|\Lambda|} \in 2m\Delta_0} dp \int_{\substack{\frac{U_1(q)}{|\Lambda|} \in \Delta_1 \\ q \in \Delta^N}} \omega_N(dq)$$

The p-integral is

$$\frac{(2\pi m|\Lambda|u_0'')^{\frac{Nd}{2}} - (2\pi m|\Lambda|u_0')^{\frac{Nd}{2}}}{(\frac{Nd}{2})!} \quad \text{if} \quad \Delta_0 = (u_0',u_0'') ,$$

so with sufficient accuracy it is equal to $(\frac{2\pi m|\Lambda|u_0''}{\frac{Nd}{2e}})^{\frac{Nd}{2}}$, and $\Omega_\Lambda(A)$

is approximatively equal to

$$\Omega_\Lambda(A) \approx (\frac{4\pi meu_0}{du_2})^{\frac{du_2|\Lambda|}{2}} \bar{\Omega}_\Lambda(\Delta_1 \times \Delta_2) ,$$

so when $|\Lambda| \to \infty$, and then $A \to (u_0,u_1,u_2)$ we get

$$s(u_0,u_1,u_2) = \frac{du_2}{2} \log \left(\frac{4\pi m e u_0}{du_2}\right) + \bar{s}(u_1,u_2) .$$

Let us call the first part $\bar{\bar{s}}(u_0,u_2)$, the "kinetic entropy".

The entropy $s(e,n)$ of the total system described by $H(p,q) = U_0(p) + U_1(q)$, and $N = U_2$ can now be identified from

$s(A) = \sup\limits_{(e,n)\in A} s(e,n)$ for $A \subset R^2$, Indeed, $(e,n) \in A$ corresponds to

$(u_0 + u_1, u_2) \in A$, so

$$s(A) = \sup\limits_{(u_0+u_1,u_2)\in A} s(u_0,u_1,u_2) = \sup\limits_{(e,n)\in A}\ (\sup\limits_{u_0} s(u_0,e-u_0,n)) ,$$

and hence

$$s(e,n) = \sup\limits_{u_0} s(u_0,e-u_0,n) = \sup\limits_{u_0} \bar{\bar{s}}(u_0,n) + \bar{s}(e-u_0,n)$$

if this is an u.s.c. function of (e,n) according to Lemma 5 c). We have to check that if $s_i \leq s(e_i,n_i)$ and $(s_i,e_i,n_i) \rightarrow (s,e,n)$ then $s \leq s(e,n)$. Note that $\bar{\bar{s}}(u,n) = -\infty$ for $u < 0$ and $\bar{s}(e-u,n) = -\infty$ for $e-u < -Kn$ because $U_1(q)$ is stable, so in the $\sup\limits_{u} \bar{\bar{s}}(u,n) + \bar{s}(e-u,n)$ only values in the compact inverval $0 \leq u \leq e + Kn$ take part. Hence $\sup\limits_{u}$ is actually

attained for some u in this interval because $\bar{\bar{s}}(u,n)$ and $\bar{s}(e-u,n)$ are u.s.c. Let $s(e_i,n_i) = \bar{\bar{s}}(u_i,n_i) + \bar{s}(e_i-u_i,n_i)$, and by passing to a subsequence suppose that $u_i \rightarrow u$. Then

$$s \leq \lim\limits_{i} \bar{\bar{s}}(u_i,n_i) + \lim\limits_{i} \bar{s}(e_i-u_i,n_i) \leq$$

$$\leq \bar{\bar{s}}(u,n) + \bar{s}(e-u,n) \leq s(e,n)$$

as claimed. $\bar{\bar{s}}(u,n)$ is increasing and differentiable in u, and these properties are inherited by $s(e,n)$ as a function of e. Indeed, suppose $e' > e''$ and $s(e',n) = \bar{\bar{s}}(e'-u',n) + \bar{s}(u',n)$ etc. Then $s(e',n) \geq \bar{\bar{s}}(e'-u'',n) + \bar{s}(u'',n) > \bar{\bar{s}}(e''-u'',n) + \bar{s}(u'',n) = s(e'',n)$.

From Lemma 7 concerning conjugate functions we recall that a convex function is differentiable iff its conjugate is strictly convex. This is easy to check for $s(\cdot,n)$. In fact its conjugate is:

$$h(\beta,n) = \sup_{e} (s(e,n) - \beta e) =$$

$$= \sup_{e,u} (\bar{\bar{s}}(u,n) + \bar{s}(e-u,n) - \beta u - \beta(e-u)) =$$

$$= \sup_{u} (\bar{\bar{s}}(u,n) - \beta u) + \sup_{e} (\bar{s}(e,n) - \beta e) =$$

$$= \bar{\bar{h}}(\beta,n) + \bar{h}(\beta,n) \ ,$$

and $\bar{\bar{h}}(\beta,n)$ is strictly convex in β because $\bar{\bar{s}}(\cdot,n)$ is differentiable, so the same is true for $h(\beta,n)$.

$\left(\bar{\bar{h}}(\beta,n) \text{ is defined by } \bar{\bar{s}}_{u}(u,n) = \beta \ , \quad \text{i.e.}\right.$

$\dfrac{dn}{2u} = \beta \ , \ u = \dfrac{dn}{2\beta} \ , \ \bar{\bar{h}}(\beta,n) = \dfrac{dn}{2} \log \left(\dfrac{2\pi me}{\beta}\right) - \dfrac{dn}{2} = \dfrac{nd}{2} \log \left(\dfrac{2\pi m}{\beta}\right)\Big)$

Hence we see that it is the presence of the kinetic energy which accounts for the physically important fact that $s(\cdot,n)$ is increasing and β positive.

We can now also prove that $s(e,n)$ is differentiable as a function of (e,n) . Lemma 7 tells us that this happens iff there is a unique supporting plane at each point (e,n), and since $s'_e(e,n)$ exists this happens iff $s'_n(e,n)$ exists, because such a plane is determined if its slopes in two directions are given. We now claim that $s'_n(e,n)$ exists iff $g(\beta,\mu)$ is strictly convex in $\beta\mu$. In fact, as a function of $\beta\mu$ $g(\beta,\mu)$ is the conjugate of $h(\beta,n)$:

$$g(\beta,\mu) = \sup_{e,n} (s(e,n) - \beta e + \beta\mu n) = \sup_{n}(h(\beta,n) + \beta\mu n),$$

so Lemma 7 tells us that $g(\beta,\cdot)$ is strictly convex if $h(\beta,\cdot)$ is differentiable. (Check that $h(\beta,\cdot) = \bar{\bar{h}}(\beta,\cdot) + \bar{h}(\beta,\cdot)$ is convex, u.s.c.,

$\bar{\bar{h}}(\beta,n) = \dfrac{nd}{2} \log \left(\dfrac{2\pi m}{\beta}\right)$, so it is enough to check $\bar{h}(\beta,n) = \sup_{e \geq -Kn} (\bar{s}(e,n) - \beta e)$.

The sup is always attained because $\bar{s}(e,n) - \beta e \leq 1 - \beta e \to -\infty$ as $e \to +\infty$. If $\bar{h}_i \leq \bar{h}(\beta,n_i) = \bar{s}(e_i,n_i) - \beta e_i$ and $(\bar{h}_i,n_i) \to (\bar{h},n)$, then $-Kn_i \leq e_i \leq \beta^{-1}(1 - \bar{h}_i)$, so we can suppose that $e_i \to e$ and u.s. continuity follows as for $s(e,n)$ above. Convexity is easily checked: if $n = \lambda n' + (1-\lambda)n''$, $\bar{h}(\beta,n') = \bar{s}(e',n') - \beta e'$ etc. then if $e = \lambda e' + (1-\lambda)e''$

$$\bar{h}(\beta,n) \geq \bar{s}(e,n) - \beta e \geq \lambda(\bar{s}(e',n') - \beta e') + (1 - \lambda)(\bar{s}(e'',n'') - \beta e'') =$$

$$= \lambda\bar{h}(\beta,n') + (1 - \lambda)\bar{h}(\beta,n'').)$$

(Remark: Also

$$h(\beta,n) = \sup_{e} (s(e,n) - \beta e) =$$

$$= \sup_{u} (\bar{\bar{s}}(u,n) - \beta u) + \sup_{e-u} (\bar{s}(e-u,n) - \beta(e-u))$$

is always attained, because we have just seen that the last sup is, and
the first one is for $u = \frac{nd}{2\beta}$. This fact will soon be needed.)

Moreover $h(\beta,\cdot)$ is differentiable iff there is a unique supporting line
(with slope $-\beta\mu$) for each n. $-\beta\mu$ is such a slope iff

$$g(\beta,\mu) = \sup_{n} (h(\beta,n) + \beta\mu n) = h(\beta,n) + \beta\mu n = s(e,n) - \beta e + \beta\mu n \quad \text{for some } e,$$

i.e. iff $(\beta,-\beta\mu)$ defines a supporting plane to s at (e,n), and such
a plane is unique as remarked above iff $s_n'(e,n)$ is defined, and
$\beta\mu = - s_n'(e,n)$.

Hence it now remains to prove that $g(\beta,\mu)$ is strictly convex in $\beta\mu$.
If it is not there is an interval (μ',μ'') where it is linear, i.e.
where $g_\mu'(\beta,\mu)$ exists and is constant. From Lemma 9 follows that then
$$\frac{\partial g(\beta,\mu)}{\partial\mu} = \lim_{\Lambda} \frac{\partial g_\Lambda(\beta,\mu)}{\partial\mu} , \quad \text{so} \quad 0 = g_\mu'(\beta,\mu_2) - g_\mu'(\beta,\mu_1) = \lim_{\Lambda} \int_{\mu_1}^{\mu_2} \frac{\partial^2 g_\Lambda(\beta,\mu)}{\partial\mu^2} d\mu$$

for $\mu' < \mu_1 < \mu_2 < \mu''$.

Hence, if we can show that $\dfrac{\partial^2 g_\Lambda(\beta,\mu)}{\partial\mu^2} \geq c > 0$ uniformly in the vicinity

of each point (β,μ) it follows that $g(\beta,\mu)$ is strictly convex in $\beta\mu$.
Let Λ be a big cube, Λ_ℓ e.g., which can be partitioned into $K = 2^{\ell d}$
cubes with side L. Take e.g. $L > 4R$. As we have seen several times

before $\dfrac{\partial^2 g_\Lambda(\beta,\mu)}{\partial(\beta\mu)^2} = \dfrac{Var(N)}{|\Lambda|}$ in the g. can. ensemble in Λ defined by

(β,μ), so we have to show that $Var(N) \geq c|\Lambda| > 0$. Let $\{\Lambda_i\}_1^K$ be the

cubes of side L making up Λ, and consider Λ as $\Lambda_0 \bigcup_1^K \Lambda_i'$,

$\Lambda_0 = \Lambda \setminus \bigcup_1^K \Lambda_i'$. Any configuration $x = (p,q)$ in Λ is correspondingly partitioned into $(x_0, x_1, \ldots x_K)$ depending on whether $q_i \in \Lambda_0$ or Λ_i' , and similarly $N = \sum_0^K N_i$. Given x_0 $\{x_i\}_1^K$ and $\{N_i\}_1^K$ are independent since $\{q_i\}_1^K$ have no mutual interaction, only interaction with q_0 .

The general relation for random variables X, Y:

$$\mathrm{Var}(X) = E\,\mathrm{Var}(X|Y) + \mathrm{Var}\,E(X|Y) \geq E\,\mathrm{Var}(X|Y)$$

then gives $\mathrm{Var}(N) \geq E\,\mathrm{Var}(N|x_0) = \sum_1^K E\,\mathrm{Var}(N_i|x_0)$.

The distribution of N_i given x_0 is determined by the g. can. distribution in Λ_i' with interaction energy $H(q_i|q_0)$ between q_i and that part of q_0 which is in $\Lambda_i \setminus \Lambda_i'$. Hence

$$p_n(x_0) = p(N_i = n|x_0) = \frac{\displaystyle\int_{x \in R^{nd} \times (\Lambda_i')^n} e^{-\beta H(x) - \beta H(q|q_0) + \beta\mu n} \, dp\omega_n(dq)}{\displaystyle\sum_n \ \frac{}{} \ \text{d:o}}$$

If we can bound e.g. $p_0(x_0)$ and $p_1(x_0)$ from below by e.g. $p(x_0) > 0$ then we can also bound $\mathrm{Var}(N_i|x_0)$ from below:

Put $Y = \begin{cases} 1 & \text{if} \quad N_i \leq 1 \\ 0 & \text{if} \quad N_i > 1 \end{cases}$.

Then

$$\mathrm{Var}(N_i|x_0) \geq E\,\mathrm{Var}(N_i|x_0, Y) \geq$$

$$\geq P(Y = 1) \cdot \mathrm{Var}(N_i|x_0, Y=1) = (p_0 + p_1)\,\frac{p_0 p_1}{(p_0 + p_1)^2} = \frac{1}{p_0^{-1} + p_1^{-1}} \geq \frac{p(x_0)}{2} \ .$$

To find $p(x_0)$ suppose that the interaction energy has the bound:
$H(q|q_0) \geq - K \cdot N_i \cdot M_i$ for some $K > 0$, M_i being the number of particles in $\Lambda_i \setminus \Lambda_i'$ interacting with Λ_i' . Such a bound is valid if there is only a pair interaction bounded below by $-K$ e.g. We can then bound $p_0(x_0)$:

$$\frac{1}{p_0(x_0)} = \sum_{n \geq 0} \int_{(p,q) \in R^{nd} \times (\Lambda_i')^n} e^{-\beta H(x) - \beta H(q|q_0) + \beta \mu n} \frac{dpdq}{n!} \leq$$

$$\leq \sum_{n \geq 0} \int_{p \in R^{nd}} e^{-\frac{\beta|p|^2}{2m}} dp \; e^{\beta K n + \beta K n M_i + \beta \mu n} \frac{|\Lambda_i'|^n}{n!} =$$

$$= \exp\left(\frac{2\pi m}{\beta}\right)^{\frac{d}{2}} e^{\beta K + \beta K M_i + \beta \mu} (L - 2R)^d \leq G_M$$

if $M_i \leq M$ and (β,μ) vary in a neighbourhood of fixed values. Similarly:

$$p_1(x_0) \geq \frac{1}{G_M} \int_{(p,q) \in R^d \times (\Lambda_i')'} e^{-\beta H(x) + \beta \mu} dpdq = \frac{\left(\frac{2\pi m}{\beta}\right)^{\frac{d}{2}} e^{\beta \mu} (L - 4R)^d}{G_M},$$

because if we restrict the particle to be in $(\Lambda_i')'$ it does not interact at all with q_0 .

Hence we have a bound $p(x_0) \geq 2q_M > 0$ if $M_1 \leq M$ and (β,μ) vary near fixed values.

Let K_M be the number of Λ_i which have $M_i \leq M$. Then we see that $\text{Var}(N) \geq E(K_M \cdot q_M) = q_M \cdot E(K_M)$. The remaining $K - K_M$ cubes have $M_i > M$, so $(K - K_M) \cdot M \leq \sum_i M_i \leq N$, and

$$K_M \geq K - \frac{N}{M} , \quad E(K_M) \geq K - \frac{1}{M} E(N) .$$

We have $\dfrac{E(N)}{|\Lambda|} = \dfrac{\partial g_\Lambda(\beta,\mu)}{\partial(\beta\mu)} \leq \dfrac{g_\Lambda(\beta,\mu') - g_\Lambda(\beta,\mu)}{\beta(\mu' - \mu)}$

for any $\mu' > \mu$ by the convexity of $g_\Lambda(\beta,\mu)$.

Hence $\dfrac{E(N)}{|\Lambda|}$ is bounded as $\Lambda \to \infty$ because $g_\Lambda(\beta,\mu) \to g(\beta,\mu)$. $K \cdot L^d = |\Lambda|$, so we finally see that

$$\text{Var}(N) \geq q_M \left(|\Lambda| L^{-d} - \frac{\text{const}}{M} |\Lambda|\right) = |\Lambda| \; q_M (L^{-d} - \frac{\text{const}}{M}) = |\Lambda| \; c > 0$$

if M is chosen big enough.

Let us finally study the convex region $D \subset R^2$ where $s(e,n) > -\infty$. We
show that this happens in the interior of a region of the following shape:

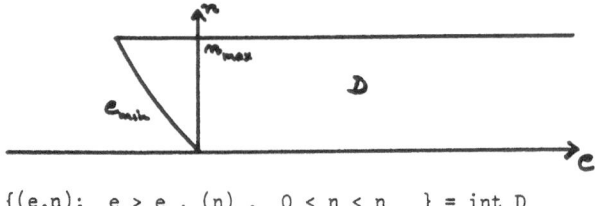

$$\{(e,n); \quad e > e_{min}(n), \quad 0 < n < n_{max}\} = int\ D$$

$e_{min}(n) \geq -Kn$ is a concave function of n.

In fact: $s(e,n) = \sup_u (\bar{s}(u,n) + \bar{s}(e-u,n))$, so if $\bar{s}(e,n) > -\infty$ for some
$e \geq e_{min}(n)$ then as soon as $e > e_{min}(n)$ we can find $u > 0$ so that
$e - u \geq e_{min}(n)$, i.e. so that $\bar{s}(u,n) + \bar{s}(e-u,n) > -\infty$. We hence have
to show that $\sup_e \bar{s}(e,n) > -\infty$ in some interval $0 < n < n_{max}$, and can

then take $e_{min}(n) = \inf_{\bar{s}(e,n)>-\infty} e$ for these n-values.

If we put no restriction on $U_1(q)$ in the definition of $\bar{s}(A)$, i.e. if
$A = R^1 \times \Delta$, with $\Delta = (n',n'')$ then

$$\Omega_\Lambda(A) = \sum_{\frac{N}{|\Lambda|} \in \Delta} \int_{q \in \Lambda^N} \omega_N(dq)$$

If there are no hard core restrictions between the particles then the
integral is

$$\frac{|\Lambda|^N}{N!} \geq (\frac{|\Lambda|}{N})^N, \quad \text{and} \quad \Omega_\Lambda(A) \geq |\Lambda|(n''-n') \inf_{n \in \Delta} (\frac{1}{n})^{n|\Lambda|}$$

so that we get:

$$\bar{s}(A) = \sup_{n \in \Delta} \sup_e \bar{s}(e,n) \geq \inf_{n \in \Delta} (n \log \frac{1}{n}) > -\infty$$

for any finite interval Δ.

For any $n > 0$ we can hence find $n_1 < n < n_2$ with
$\sup_e \bar{s}(e,n_1)$, $\sup_e \bar{s}(e,n_2) > -\infty$ by taking Δ below or above n.

If $n = \lambda n_1 + (1-\lambda)n_2$, $\bar{s}(e_i,n_i) > -\infty$, $e = \lambda e_1 + (1-\lambda)e_2$ then we see
that $\sup_e \bar{s}(e,n) \geq \bar{s}(e,n) \geq \lambda\bar{s}(e_1,n_1) + (1-\lambda)\bar{s}(e_2,n_2) > -\infty$.

I.e. $n_{max} = +\infty$ in this case.

If there are hard core restrictions so that $|q_i - q_j| \geq r > 0$ for all particles then n_{max} is at most equal to the close packing density of spheres. To see that $n_{max} > 0$ we remark that a lower bound to $\Omega_\Lambda(A)$ in this case is obtained if Λ is a cube which contains N cubes of side L regularly spaced with spacing $L + 2r$:

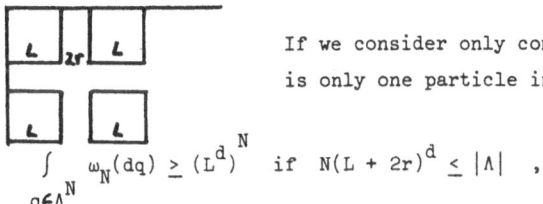

If we consider only configurations where there is only one particle in each small cube we see that

$$\int_{q\in\Lambda^N} \omega_N(dq) \geq (L^d)^N \quad \text{if} \quad N(L + 2r)^d \leq |\Lambda| \quad ,$$

and as before

$$\bar{s}(A) = \inf_{n\in A} n \log L^d > - \infty \quad \text{if} \quad n \leq (L + 2r)^{-d} .$$

The best choice of L is $L = n^{-1/d} - 2r$, so we have

$$\bar{s}(A) \geq \inf_{n\in A} nd \log(n^{-1/d} - 2r) > - \infty \quad \text{if} \quad n" < (2r)^{-d} , \quad \text{and as before}$$

$$\sup_{e} s(e,n) > - \infty \quad \text{for} \quad 0 < n < (2r)^{-d} , \quad \text{so} \quad n_{max} \geq (2r)^{-d} .$$

By a similar argument we can see that $\bar{s}(0,0) = 0 > - \infty$ so that $\bar{s}(e,0) > - \infty$ also for all $e \geq 0$.

In fact let $A = A_1 \times A_2$ be a neighbourhood of $(0,0)$ and consider the restricted configurations above but with r changed to R. All of them have $\dfrac{U_1(q)}{|\Lambda|} = 0 \in A_1$, so they are allowed, and again we have

$$\bar{s}(A) \geq \inf_{n\in A_2} nd \log (n^{-1/d} - 2R), \quad \text{and as} \quad A \quad \text{shrinks to} \quad (0,0): \bar{s}(0,0) \geq 0.$$

An upper bound to $\bar{s}(A)$ is obtained by ignoring the restriction on $U_1(q)$:

$$\Omega_\Lambda(A) \leq \sum_{\frac{N}{|\Lambda|}\in A_2} \frac{|\Lambda|^N}{N!} \leq \sum_{\frac{N}{|\Lambda|}\in A_2} \left(\frac{e|\Lambda|}{N}\right)^N ,$$

which gives $\bar{s}(A) \leq \sup_{n\in A_2} n \log \left(\frac{e}{n}\right)$, and as A shrinks to $(0,0)$:

$$\bar{s}(0,0) \leq 0.$$

Let us now collect the properties of $s(e,n)$ we have found:

Theorem 6. Let $H(p,q) = \sum_{1}^{N} \frac{|p_i|^2}{2m} + U_1(q) = U_0(p) + U_1(q)$, where

$U_1(q)$ is stable, $U_1(q) \geq K \cdot N$, and has finite range R. Suppose also that for any two configurations q_1, q_2 with N_1, N_2 particles the mutual interaction $U_1(q_1, q_2) - U_1(q_1) - U_1(q_2) \geq - K \cdot N_1 N_2$. Then the entropy for the observables (H,N) is $s(e,n) = \sup_{0 \leq u \leq e + Kn} \overline{\overline{s}}(u,n) + \overline{s}(e-u,n)$, where

$\overline{\overline{s}}(u,n) = \frac{nd}{2} \log (\frac{4\pi meu}{nd})$ is the entropy for (U_0, N) and $\overline{s}(e,n)$ the entropy

for (U_1, N). $s(e,n) > - \infty$ in $D \subset R^2$, where

$\quad$ int $D = \{(e,n) ; e > e_{min}(n) , 0 < n < n_{max}\}$.

$s(e,n)$ is increasing in e and differentiable in (e,n) , and bounded, $s(e,n) \leq \overline{\overline{s}}(e + Kn, n) + 1$.

The conjugate function $g(\beta, \mu) = \sup_{e,n} (s(e,n) - \beta e + \beta \mu n)$ is bounded:

$0 \leq g(\beta, \mu) \leq (\frac{2\pi m}{\beta})^{\frac{d}{2}} e^{\beta K + \beta \mu}$ $\quad$ for $\beta > 0$, μ arbitrary and strictly convex

in $\beta \mu$.

3.4.2. The existence of s(e,n) when the interaction has infinite range

In this section we show how the proof of Theorem 2 can be modified when the interaction is of infinite range. For simplicity we only consider the case $U(q) = (U_1(q),N)$, $a = (\beta,-\beta\mu)$, i.e we have only potential energy (and omit the bar used in the notation in the previous section). We have to make some assumption about the decay of the interaction energy between configurations far apart however, and a useful one is the following:

Definition: $U_1(q)$ is called tempered if for some $R>0$ and $\delta>d$

$$|U_1(q_1,q_2) - U_1(q_1) - U_1(q_2)| \leq K \cdot N_1 \cdot N_2 (d(q_1,q_2))^{-\delta}$$

for any two configurations q_1,q_2 with N_1,N_2 particles respectively when their distance $d(q_1,q_2) \geq R$.

A pair interaction $U_1(q) = \sum_{i<j} u(q_i-q_j)$ is tempered if

$$|u(x)| \leq K \cdot |x|^{-\delta} \quad \text{for } |x| > R \quad x \in R^d.$$

We assume that $U_1(q)$ is stable and tempered. As before we first consider the special cubes Λ_ℓ with sides $L \cdot 2^\ell$ and define $\Omega'_\ell(A,a)$ by restricting all the particles to be in $\Lambda'_\ell \subset \Lambda_\ell$, where Λ'_ℓ is a cube with side $L \cdot 2^\ell - 2R_\ell$. We shall let $R_\ell \to \infty$, but more slowly than $L \cdot 2^\ell$. We also make the restriction

$$\left(\frac{U_1(q)}{|\Lambda|}, \frac{N}{|\Lambda|}\right) \in A$$

more restrictive by shrinking $A \subset R^2$ by an amount $\varepsilon_\ell \to 0$ in the e-direction.

Hence

$$\Omega'_\ell(A,a) = \int e^{-\beta U_1(q) + \beta\mu N} \omega(dq)$$

$$\left(\frac{U_1}{|\Lambda_\ell|} + \varepsilon_\ell, \frac{N}{|\Lambda_\ell|}\right) \in A$$

$$q \in \Gamma_{\Lambda'_\ell}$$

To start with we shall assume that A is bounded in the n-direction: $n \leq c$ when $(e,n) \in A$, and we shall see that we can take $R_\ell = R_0 2^{\rho \ell}$, $\varepsilon_\ell = \varepsilon_0 2^{-\varepsilon\ell}$ for suitable R_0, ε_0, $\varepsilon > 0$, $0 < \rho < 1$.

The temperedness implies that if we have several configurations $(q_1, q_2 \ldots)$ with $(N_1 N_2 \ldots)$ particles and $d(q_i, q_j) \geq D$ then if $q = (q_1, q_2 \ldots)$

$$|U_1(q) - \sum_i U_1(q_i)| \leq KD^{-\delta}(\sum_{i<j} N_i N_j) \leq KD^{-\delta}(\sum_i N_i)^2.$$

In comparing $\Omega_\ell(A,a)$ and $\Omega_{\ell+1}(A,a)$ we note that in $\Lambda'_{\ell+1}$ we can pack 2^d translates of Λ_ℓ so that their distances are at least

$$(L \cdot 2^{\ell+1} - 2R_{\ell+1}) - 2(L \cdot 2^\ell - 2R_\ell) = 4R_\ell - 2R_{\ell+1} = 2R_0 2^{\rho \ell}(2 - 2^\rho) \geq$$

$$\geq R \cdot 2^{\rho \ell} \quad \text{if } 2R_0(2 - 2^\rho) \geq R.$$

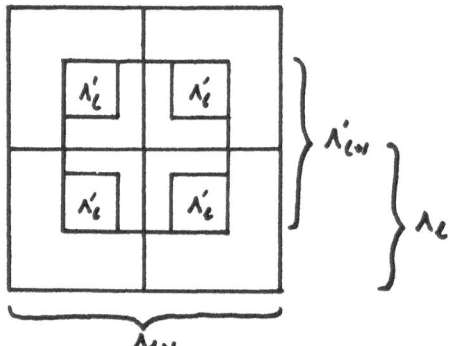

Hence if in $\Lambda'_{\ell+1}$ we consider only configurations $q = (q^{(1)}, \ldots q^{(2^d)})$ with $q^{(i)}$ in these translates of Λ'_ℓ and

$$\left(\frac{U_1(q^{(i)})}{|\Lambda_\ell|} \pm \varepsilon_\ell, \frac{N_\ell}{|\Lambda_\ell|} \right) \in A \quad \text{then}$$

$$|U_1(q) - \sum_i U_1(q^{(i)})| \leq K(c|\Lambda_{\ell+1}|)^2 (R_0 \cdot 2^{\rho \ell})^{-\delta} \equiv \Delta \quad \text{and}$$

$$\frac{U_1(q)}{|\Lambda_{\ell+1}|} + \varepsilon_\ell \leq 2^{-d} \sum_i \left(\frac{U_1(q^{(i)})}{|\Lambda_\ell|} + \varepsilon_\ell \right) + \frac{\Delta}{|\Lambda_{\ell+1}|}$$

$$\frac{U_1(q)}{|\Lambda_{\ell+1}|} - \varepsilon_\ell \geq 2^{-d} \sum_i \left(\frac{U_1(q^{(i)})}{|\Lambda_\ell|} - \varepsilon_\ell \right) - \frac{\Delta}{|\Lambda_{\ell+1}|} \quad ,$$

so that by the convexity of A

$$\left(\frac{U_1(q)}{|\Lambda_{\ell+1}|} \pm (\varepsilon_\ell - \frac{\Delta}{|\Lambda_{\ell+1}|}), \frac{N}{|\Lambda_{\ell+1}|} \right) \in A. \quad (N = \sum_i N^{(i)}.)$$

Hence if $\varepsilon_{\ell+1} \leq \varepsilon_\ell - \dfrac{\Delta}{|\Lambda_{\ell+1}|}$ we have

$$\left(\frac{U_1(q)}{|\Lambda_{\ell+1}|} \pm \varepsilon_{\ell+1} , \frac{N}{|\Lambda_{\ell+1}|}\right) \in A,$$

so q is allowed in $\Lambda_{\ell+1}$ and as before

$$\Omega'_{\ell+1}(A,a) \geq (\Omega'_\ell(A,a))^{2^d} \cdot e^{-\beta\Delta},$$

so for $s_\ell = \dfrac{1}{|\Lambda_\ell|} \log \Omega_\ell$ we have

$$s_{\ell+1} \geq s_\ell - \beta \frac{\Delta}{|\Lambda_{\ell+1}|} \geq s_\ell + \beta(\varepsilon_{\ell+1} - \varepsilon_\ell), \text{ or}$$

$$s_{\ell+1} - \beta\varepsilon_{\ell+1} \geq s_\ell - \beta\varepsilon_\ell.$$

So this time $s_\ell - \beta\varepsilon_\ell$ is increasing, and has a limit $s(A,a)$, and since $\varepsilon_\ell \to 0$ $s_\ell \to s(A,a)$ also. Just as before it is easy to check that $s(A,a)$ is inner regular in A and has the properties of Lemma 4, so it is given by $s(A,a) = \sup_{(e,n)\in A} (s(e,n) - \beta e + \beta\mu n)$ with $s(e,n) = \inf_{A \ni (e,n)} s(A,0)$.

It remains to check that ε_ℓ can be found so that

$$\varepsilon_\ell - \varepsilon_{\ell+1} \geq \frac{\Delta}{|\Lambda_{\ell+1}|} = Kc^2 R_0^{-\delta}(2L)^d 2^{\ell(d-\rho\delta)}. \quad (\varepsilon_\ell - \varepsilon_{\ell+1}) = \varepsilon_0(1-2^{-\varepsilon})2^{-\varepsilon\ell}.$$

Hence we see that since $\delta > d$ we can take $\rho < 1$ so that still $\rho\delta > d$, and then $0 < \varepsilon \leq (\rho\delta-d)$, and finally ε_0 so that

$$\varepsilon_0(1-2^{-\varepsilon}) \geq Kc^2 R_0^{-\delta}(2L)^d.$$

In order to treat more arbitrary $\Lambda \to \infty$ we have to slightly strengthen the van Hove condition:

<u>Definition</u>: $\{\Lambda_i\}$ is said to tend to infinity in the strong van Hove sense if $|\Lambda_i| \to \infty$ and if

$$\frac{|\Lambda_i(R)|}{|\Lambda_i|} \leq f\left(\frac{R^d}{|\Lambda_i|}\right), \text{ where } f(x) \text{ is a continuous decreasing function}$$

for $x \geq 0$ with $f(0) = 0$.

The proof of Lemma 11 then goes roughly as before:

<u>Lemma 11'</u> If $\Lambda \to \infty$ in the strong van Hove sense and if $\Lambda' \subset \Lambda$ is defined by deleting a corridor of width R_Λ along the boundary of Λ then if $\dfrac{R_\Lambda^d}{|\Lambda|} \to 0$ and if $\Omega'_\Lambda(A,a)$ is defined by confining the particles to Λ'

$$\lim_{\Lambda \to \infty} \frac{1}{|\Lambda|} \log \Omega_\Lambda(A,a) \geq \lim_{\Lambda \to \infty} \frac{1}{|\Lambda|} \log \Omega'_\Lambda(A,a) \geq s(A,a)$$

if A is bounded in the n-direction.

<u>Proof</u>: As before we fill Λ' with N_ℓ translates of Λ_ℓ, $\{\Lambda_i\}_1^{N_\ell}$ so that if $\quad \Lambda_0 = \bigcup_1^{N_\ell} \Lambda_i$

$$|\Lambda'| - |\Lambda_0| \leq |\Lambda(D+R)| \quad , \quad |\Lambda| - |\Lambda'| \leq |\Lambda(R)| .$$

Take $C \subset A$ and consider only configurations $q = (q_1, q_2, \ldots)$ with $q_i \in \Lambda'_i$, Λ'_i being centered in Λ_i with side $L \cdot 2^\ell - 2R_\ell$, R_ℓ as before. Then by the temperedness

$$|U(q) - \sum_i U(q_i)| \leq Kc^2 (N_\ell |\Lambda_\ell|)^2 (R_0 2^{\rho \ell})^{-\delta} \equiv \Delta$$

$$\left| \frac{U(q)}{|\Lambda|} - \frac{N_\ell |\Lambda_\ell|}{|\Lambda|} \frac{1}{N_\ell} \sum_1^{N_\ell} \frac{U(q_i)}{|\Lambda_\ell|} \right| \leq \frac{\Delta}{|\Lambda|}$$

Hence if $\dfrac{\Delta}{|\Lambda|} \to 0$, $\dfrac{N_\ell |\Lambda_\ell|}{|\Lambda|} \to 1$ we see that if $\left(\dfrac{U(q_i)}{|\Lambda_\ell|}, \dfrac{N_i}{|\Lambda_\ell|} \right) \in C$ for all i then $\left(\dfrac{U(q)}{|\Lambda|}, \dfrac{N}{|\Lambda|} \right) \in A$ if Λ is big enough so that q is allowed in Λ',

and as before we have

$$\Omega'_\Lambda(A,a) \geq (\Omega'_\ell(C,a))^{N_\ell} e^{-\beta \Delta}$$

$$\frac{1}{|\Lambda|} \log \Omega'_\Lambda(A,a) \geq \frac{N_\ell |\Lambda_\ell|}{|\Lambda|} \frac{1}{|\Lambda_\ell|} \log \Omega'_\ell(C,a) - \beta \frac{\Delta}{|\Lambda|} .$$

Hence if we can arrange so that $\Lambda \to \infty, \ell \to \infty$ together so that $\dfrac{\Delta}{|\Lambda|} \to 0$, $\dfrac{N_\ell |\Lambda_\ell|}{|\Lambda|} \to 1$ we get $\lim_\Lambda \geq s(C,a)$ for all $C \subset A$ and hence $\lim_\Lambda \geq s(A,a)$.

We need to have $\dfrac{D^d}{|\Lambda|} \to 0$, i.e. $\dfrac{2^{\ell d}}{|\Lambda|} \to 0$ in addition to $\dfrac{R^d}{|\Lambda|} \to 0$ in order that $\dfrac{N_\ell|\Lambda_\ell|}{|\Lambda|} \to 1$ as before. $\dfrac{\Delta}{|\Lambda|} \le Kc^2 R_0^{-\delta}|\Lambda|2^{-\ell\rho\delta}$, so we need to have $|\Lambda|2^{-\ell\rho\delta} \to 0$, and $|\Lambda|2^{-\ell d} \to \infty$. This is achieved if e.g. ℓ is chosen so that $|\Lambda| \approx 2^{(\rho\delta+d)\ell/2}$ as $|\Lambda| \to \infty$.

Lemma 12' obtained by requiring that $\Lambda \to \infty$ in the strong van Hove sense and that n is bounded in A can then be proved roughly as Lemma 12.

Proof: $\Lambda_1 \supset \Lambda$ and Λ_2 are defined as before, and Λ_2' is obtained from Λ_2 by deleting a corridor of width R_ℓ along $\partial\Lambda_2$, Λ_1 having side $L \cdot 2^\ell$. We take $C \subset A$ and consider only configurations $q_1 = (q, q_2)$ in $\Lambda \cup \Lambda_2'$ with restrictions defined by A and C respectively. Then

$$|U_1(q_1) - U_1(q) - U_1(q_2)| \le Kc^2|\Lambda_1|^2(R_0 2^{\rho\ell})^{-\delta} \equiv \Delta$$

$$\left|\frac{U_1(q_1)}{|\Lambda_1|} - \frac{|\Lambda|}{|\Lambda_1|}\frac{U_1(q)}{|\Lambda|} - \frac{|\Lambda_2|}{|\Lambda_1|}\frac{U_1(q_2)}{|\Lambda_2|}\right| \le \frac{\Delta}{|\Lambda_1|}.$$

Since $\dfrac{|\Lambda_2|}{|\Lambda_1|} \ge \dfrac{1}{2}$ the sum of the two energies is contained in

$$\{\lambda u_2 + (1-\lambda)u; \; u_2 \in C, \; u \in A, \; \lambda \ge 1/2\},$$

and since $\bar{C}$ is compact $\subset A$ this set has a positive distance to A^c. Hence since $\dfrac{\Delta}{|\Lambda_1|} \to 0$ we can conclude that

$$\frac{U_1(q_1)}{|\Lambda_1|} \pm \varepsilon_\ell \in A$$

if ℓ is big enough so that q_1 is allowed in the definition of $\Omega_\ell'(A,a)$, and as before $\Omega_\Lambda(A,a)\Omega_{\Lambda_2'}(C,a)e^{-\beta\Delta} \le \Omega_\ell'(A,a)$.

Λ_2 also tends to infinity in the strong van Hove sense because $|\Lambda_2(R)| \le |\Lambda(R)| + |\Lambda_1(R)|$ and $\dfrac{1}{2}|\Lambda_1| \ge |\Lambda| \ge c|\Lambda_1|$,

$(1-c)|\Lambda_1| \ge |\Lambda_2| \ge \dfrac{1}{2}|\Lambda_1|$, so

$$\frac{|\Lambda_2(R)|}{|\Lambda_2|} \le \frac{|\Lambda(R)|}{|\Lambda|} + 2\frac{|\Lambda_1(R)|}{|\Lambda_1|} \le f(\frac{R^d}{|\Lambda|}) + \text{const.}\frac{R}{|\Lambda|^{1/d}} \le$$

$$\le f(\frac{1-c}{c}\frac{R^d}{|\Lambda_2|}) + \text{const.}\,(\frac{R^d}{|\Lambda_2|})^{1/d} \equiv f_2(\frac{R^d}{|\Lambda_2|}) \to 0 \text{ as } \frac{R^d}{|\Lambda_2|} \to 0.$$

Now $\dfrac{R_\ell^d}{|\Lambda_2|} \le \text{const.}\, 2^{\ell d(\rho-1)} \to 0$, and $\dfrac{\Delta}{|\Lambda_1|} \to 0$, so as before we can apply

Lemma 11$'$ to Ω_{Λ_2}' and get

$$c \, \overline{\lim_{\Lambda \to \infty}} \, \frac{1}{|\Lambda|} \log \Omega_\Lambda(A,a) + (1-c)s(C,a) \leq s(A,a)$$

for all $C \subset A$, and hence

$$\overline{\lim_\Lambda} \leq s(A,a) \quad \text{if } s(A,a) > -\infty.$$

The case $s(A,a) = -\infty$ can be treated as before.

We finally get rid of the restriction that n is bounded in A. For any open convex $A \subset R^2$ we define $s(A,a)$ by

$$s(A,a) = \sup_{(e,n) \in A} \, (s(e,n) - \beta e + \beta \mu n).$$

If $n \geq c$ in A we have:

$$\Omega_\Lambda(A,a) \leq \sum_{N \geq c|\Lambda|} \int_{q \in \Lambda^N} e^{-\beta U_1 + \beta \mu N} \omega_N(dq) \leq$$

$$\leq \sum_{c|\Lambda|}^\infty e^{(\beta K + \beta \mu)N} \frac{|\Lambda|^N}{N!} \leq (e^{\beta K + \beta \mu + 1} \frac{c|\Lambda|}{c})^{\frac{}{}} \qquad \text{if } c \geq e^{\beta K + \beta \mu}$$

(Because $\sum_b^\infty \frac{x^N}{N!} = \sum_b^\infty (\frac{x}{b})^N \frac{b^N}{N!} \leq (\frac{x}{b})^b \sum_0^\infty \frac{b^N}{N!} = (\frac{ex}{b})^b \qquad$ if $b \geq x$.)

Hence $s(A,a) \leq c(\beta K + \beta \mu + 1) - c \log c \quad$ if $n \geq c$ in A.

If $s = s(A,a) > -\infty$ take c so large that $n(\beta K + \beta \mu + 1) - n \log n \leq s-1$

e.g. for $n \geq c$. Then if A is partitioned into $A' \cup A''$ according to whether $n < c+1$ or $n > c$ respectively we have $s = \max(s(A',a), s(A'',a))$ and $s(A'') \leq s-1$, so

$$s = s(A',a) = \lim_{\Lambda \to \infty} \frac{1}{|\Lambda|} \log \Omega_\Lambda(A',a)$$

In particular $\Omega_\Lambda(A',a) \geq e^{|\Lambda|(s-1)}$ if Λ is large, and since $\Omega_\Lambda(A'',a) \leq e^{|\Lambda|(s-1)}$ by the estimate above we have

$$\Omega_\Lambda(A',a) \leq \Omega_\Lambda(A,a) \leq \Omega_\Lambda(A',a) + \Omega_\Lambda(A'',a) < 2\Omega_\Lambda(A',a), \quad \text{and hence}$$

$$\lim_{\Lambda \to \infty} \frac{1}{|\Lambda|} \log \Omega_\Lambda(A,a) = s \quad \text{also.}$$

If $s(A,a) = -\infty$ and $d(A,D) > 0$ then $s(A',a) = -\infty$ for any c, so $\Omega_\Lambda(A',a) = 0$ for Λ sufficiently large. For any s take c so that $\Omega_\Lambda(A'',a) \leq e^{|\Lambda|s}$ for all Λ. Then $\Omega_\Lambda(A,a) \leq \Omega_\Lambda(A',a) + \Omega_\Lambda(A'',a) \leq e^{|\Lambda|s}$ if Λ is sufficiently large, and

$$\varlimsup_{\Lambda \to \infty} \frac{1}{|\Lambda|} \log \Omega_\Lambda(A,a) \leq s. \quad \text{Hence } \lim_{\Lambda \to \infty} = -\infty.$$

The kinetic energy can now be taken into account just as before, but the proof that $s(e,n)$ is differentiable requires new methods to bound $\dfrac{\mathrm{Var}(N)}{|\Lambda|}$ from below.

Let us collect the results:

Theorem 7. Let $U(q) = (U_1(q),N)$, $a = (\beta, -\beta\mu)$ where $U_1(q)$ is stable and tempered, and suppose that $\Lambda \to \infty$ in the strong van Hove sense. Then

$$\lim_{\Lambda \to \infty} \frac{1}{|\Lambda|} \log \Omega_\Lambda(A,a) = \bar{s}(A,a) = \sup_{(e,n) \in A} (\bar{s}(e,n) - \beta e + \beta\mu n)$$

with $\bar{s}(e,n) = \inf_{A \ni (e,n)} \bar{s}(A,0)$

exists if $\bar{s}(A,a) > -\infty$ or if $\bar{s}(A,a) = -\infty$ and $d(A,D) > 0$. The domain D where $\bar{s}(e,n) > -\infty$ has the shape described before:

$$\mathrm{int}\ D = \{(e,n);\ e > e_{min}\ (n),\ 0 < n < n_{max}\},$$

and $\bar{s}(e,n)$ has the bound

$$\sup_e \bar{s}(e,n) \leq n \log \left(\frac{e}{n}\right). \text{ As before}$$

$$\bar{g}(\beta,\mu) = \bar{s}(R^2,a) \text{ is bounded: } 0 \leq \bar{g}(\beta,\mu) \leq e^{\beta K + \beta\mu}.$$

3.4.3. A system in a slowly varying external field, the barometric formula

In chapter 2.3 we considered a system described by a g. can. law influenced by a slowly varying external field giving a contribution
$V(\lambda q) = \sum_i v(\lambda q_i)$ to the total energy. $v(x)$ $x \in R^d$ is a nice function,

so $V(\lambda q)$ varies on a scale λ^{-1} which is long if $\lambda \to 0$. We argued
that one can then regard the system as consisting of macroscopically
infinitesimal cells of size $\lambda^{-1}\Delta x$ in which $v(x)$ is essentially constant. The cells are however microscopically infinite, so one can hope
that their interaction can be neglected, and the total partition function
is approximatively the product of those of the cells:
$G_\lambda(\beta,v) \approx \prod_x G_{cell}(\beta,-v(x))$. Then as $\lambda \to 0$ we ought to get:

$$\lim_{\lambda \to 0} \lambda^d \log G_\lambda(\beta,v) = \lim_{\lambda \to 0} \sum_x \frac{(\Delta x)^d}{|cell|} \log G_{cell}(\beta,-v(x)) = \int g(\beta,-v(x))dx .$$

Making suitable assumptions about the interaction this argument can now be
made precise:

Theorem 8: Suppose that $U_1(q) = \sum_{i<j} u(q_i-q_j)$ is a stable pair interaction
with $u(q) \geq - K$, $|u(q)| \leq K|q|^{-\delta}$ when $|q| \geq R$ for some $\delta > d, R > 0$,
and suppose that there are hard core restrictions so that $|q_i - q_j| \geq r > 0$
always and the number of particles in any region Λ is at most $c|\Lambda|$ for
some $c > 0$. Then if $e^{-\beta v(x)}$ is Riemann integrable and $\int e^{-\beta v(x)}dx < \infty$

$$G_\lambda(\beta,v) = \int e^{-\beta(U_1(q)+V(\lambda q))} \omega(dq) \text{ is finite, and}$$

$$\lim_{\lambda \to 0} \lambda^d \log G_\lambda(\beta,v) = \int g(\beta,-v(x))dx < \infty .$$

Proof: From the stability follows that $G_\lambda(\beta,v)$ is finite:

$$G_\lambda(\beta,v) \leq \sum_{N \geq 0} e^{\beta KN} \int_{q \in R^{Nd}} e^{-\beta V(\lambda q)} \frac{dq}{N!} =$$

$$= \sum_{N \geq 0} \frac{e^{\beta KN}}{N!} (\int e^{-\beta v(x)}dx)^N \lambda^{-Nd} = \exp \lambda^{-d} e^{\beta K} \int e^{-\beta v(x)}dx , \text{ and that}$$

$$\lambda^d \log G_\lambda(\beta,v) \leq e^{\beta K} \int e^{-\beta v(x)}dx .$$

The same bound is valid for $\int g(\beta, -v(x))dx$.

The following bounds hold for the interaction energies since the density is at most c: The interaction between one particle at the origin and those at a distance $\geq D \geq R$ is bounded by

$$\left| \sum_{|q_i| \geq D} u(q_i) \right| \leq \text{const.} \sum_{D}^{\infty} \frac{(\ell+1)^d - \ell^d}{\ell^\delta} \leq \text{const.} \int_{D}^{\infty} \ell^{d-1-\delta} d\ell \leq \text{const } D^{d-\delta}$$

The interaction between one particle at the origin and all others is bounded below by

$$\sum_i u(q_i) = \sum_{|q_i| \leq R} u(q_i) + \sum_{|q_i| \geq R} u(q_i) \geq - \text{const.}(R^d + R^{d-\delta}) = - \text{const.}$$

The interaction between the particles in a cube Λ with side $L \geq R$ and those outside is bounded below by

$$\sum_{q_i \in \Lambda} \sum_{q_j \in \Lambda^c} u(q_i - q_j) = \sum_{d(q_i, \Lambda^c) \leq R} \sum + \sum_{\ell=R}^{L} \sum_{\ell \leq d(q_i, \Lambda^c) < \ell+1} \sum \geq$$

$$\geq - \text{const.}(R \cdot L^{d-1} + \sum_{\ell=R}^{L-1} \frac{(L-\ell)^d - (L-\ell-1)^d}{\ell^{\delta-d}}) \geq$$

$$\geq - \text{const.}(RL^{d-1} + L^{d-1} \int_R^L \ell^{d-\delta} d\ell) \geq - \text{const.} L^{2d-\delta}.$$

Now, let Λ_k be a cube with side L centered at $L \cdot k$, $k \in \mathbf{Z}^d$, and think of R^d as partitioned into $\bigcup_k \Lambda_k$. Let Λ_k' be concentic with Λ_k having side $L - 2D$. Finally let Λ be a big cube with side $\ell \cdot \lambda^{-d}$ consisting of a certain no. of the Λ_k's.

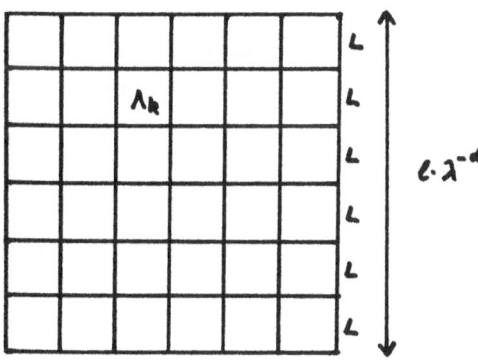

First we get a lower bound for G_λ by considering only configurations with particles only in Λ_k' for $\Lambda_k \subset \Lambda$. Then the interaction between Λ_k' and the other particles is bounded above by

$$\sum_{q_i \in \Lambda_k'} \sum_{q_j \in \Lambda_k^c} u(q_i - q_j) \leq C \cdot L^d \cdot D^{d-\delta} \equiv \Delta ,$$

so if $e^{-\beta v_k'} = \inf_{q \in \Lambda_k} e^{-\beta v(\lambda q)}$

then

$$G_\lambda(\beta, v) \geq \int_{q_k \in T_{\Lambda_k'}} e^{-\beta \Sigma(U_1(q_k) - \beta N_k v_k' - \beta \Delta)} \prod_k \omega(dq_k) = \prod_k G_{\Lambda_0'}(\beta, -v_k') e^{-\beta \Delta} ,$$

and

$$\lambda^d \log G_\lambda(\beta, v) \geq \sum_k (\lambda L)^d L^{-d} \log G_{\Lambda_0'}(\beta, -v_k') - \beta \lambda^d |\Lambda| L^{-d} \Delta .$$

As $\lambda \to 0$ and ℓ is fixed, so that $\lambda^d |\Lambda| = \ell^d$ the sum is a Riemannsum where $x = \lambda q$ ranges over the cells $\lambda \cdot \Lambda_k$ with side $\Delta x = \lambda \cdot L$. Hence we have

$$\lim_{\lambda \to 0} \lambda^d \log G_\lambda(\beta, v) \geq \int_{|x_i| \leq \ell} L^{-d} \log G_{\Lambda_0'}(\beta, -v(x)) dx - \beta \ell^d \cdot C \cdot D^{d-\delta} .$$

then as $L, D \to \infty$ with $\frac{D}{L} \to 0$ the integrand converges to $g(\beta, -v(x))$, so by bounded convergence we have

$$\lim_{\lambda \to 0} \geq \int_{|x_i| \leq \ell} g(\beta, -v(x)) dx ,$$

and finally as $\ell \to \infty$

$$\lim_{\lambda \to 0} \geq \int g(\beta, -v(x)) dx .$$

To get an upper bound we note that the previous estimates of the inter-action tell us that if $q_k \in \Gamma_{\Lambda_k}$ for $\Lambda_k \subset \Lambda$ and $q_\infty \in \Gamma_{\Lambda^c}$ and q is the total configuration then

$$U_1(q) - \sum_k U_1(q_k) - U_1(q_\infty) \geq - C \cdot L^{2d-\delta} \cdot |\Lambda| \cdot L^{-d} \equiv \Delta ,$$

so if $e^{-\beta v_k''} = \sup_{q \in \Lambda_k} e^{-\beta v(\lambda q)}$ then

$$G_\lambda(\beta,v) \leq (\prod_k G_{\Lambda_o}(\beta,-v_k'')) \; G_{\Lambda^c}(\beta,-v) \; e^{\beta\Delta} \; .$$

By the same estimate as in the beginning

$$\lambda^d \log G_{\Lambda^c}(\beta,v) \leq e^{\beta K} \cdot \int_{x \in \lambda \cdot \Lambda^c} e^{-\beta v(x)} \; dx = \epsilon_\ell \to 0$$

as ℓ, the side of $\lambda \cdot \Lambda$, $\to \infty$.

As before we thus have

$$\overline{\lim_{\lambda \to 0}} \; \lambda^d \log G_\lambda(\beta,v) \leq \int_{|x_i| \leq \ell} L^{-d} \log G_{\Lambda_o}(\beta,-v(x)) \; dx + \epsilon_\ell + \beta C L^{d-\delta} \; ,$$

and as $L \to \infty$ and then $\ell \to \infty$

$$\overline{\lim_{\lambda \to 0}} \leq \int g(\beta,-v(x)) \; dx \; .$$

3.5. The central limit theorem for macroscopic variables, thermo-dynamic fluctuation theory

When we are interested in the distribution of the macroscopic variable $U_i(q)$ in a small but macroscopic region Λ we have seen that it is given by the can. law with

$$g_\Lambda(a) = \frac{1}{|\Lambda|} \log \int_{q \in T_\Lambda} e^{-a \cdot U(q)} \; \omega(dq) \; .$$

In Theorem 3 we showed the law of large numbers for $\frac{U(q)}{|\Lambda|}$: If $u = -g'(a)$ is defined then $\frac{U(q)}{|\Lambda|} \to u$ in probability and

$$u_\Lambda = \left\langle \frac{U(q)}{|\Lambda|} \right\rangle_{\Lambda,a} = -g_\Lambda'(a) \to u \quad \text{as} \quad \Lambda \to \infty \; .$$

Since $U(q)$ is the sum of many small contributions it is not unlikely that the central limit theorem should hold for $\dfrac{U(q) - |\Lambda| u_\Lambda}{|\Lambda|^{1/2}} \equiv X(q)$.

The covariance matrix of X is $g_\Lambda''(a)$, because

$$\left\langle X_i X_j \right\rangle_{\Lambda,a} = \frac{\partial^2 g_\Lambda(a)}{\partial a_i \partial a_j} \equiv (g_\Lambda''(a))_{ij} .$$

Hence if $g_\Lambda''(a) \to g''(a)$ we can expect that the distribution of X converges to a gaussian with these covariances. This can also be seen by considering the generating function for X.

$$\left\langle e^{-b \cdot U} \right\rangle_{\Lambda,a} = \exp|\Lambda|(g_\Lambda(a+b) - g_\Lambda(a)) , \quad \text{so}$$

$$\left\langle e^{-b \cdot X} \right\rangle_{\Lambda,a} = \exp|\Lambda|(g_\Lambda(a+b|\Lambda|^{-1/2}) - g_\Lambda(a) - |\Lambda|^{-1/2} b \cdot g_\Lambda'(a)) =$$

$$= \exp|\Lambda| \int_0^1 (1-t) \frac{d^2 g_\Lambda(a+tb|\Lambda|^{-1/2})}{dt^2} \, dt =$$

$$= \exp \int_0^1 (1-t) \, b \cdot g_\Lambda''(a+tb|\Lambda|^{-1/2}) b \, dt$$

by Taylor's formula. We hence see that if $g_\Lambda''(a) \to g''(a)$ and if $\{g_\Lambda''(\cdot)\}$ are equicontinuous at a e.g. then

$$\left\langle e^{-b \cdot X} \right\rangle_{\Lambda,a} \to \exp \frac{b \cdot g''(a) b}{2}$$

and X has the gaussian limit distribution with covariance matrix $g''(a)$. Its density is usually derived in physics by the following argument: As we have seen in the proof of Theorem 3

$$P(\frac{U}{|\Lambda|} \in du) \approx \exp|\Lambda|(s(u) - a \cdot u - g(a)),$$

so if this approximation is good enough

$$P(X \in dx) \approx \exp|\Lambda|(s(u+x|\Lambda|^{-1/2}) - a \cdot (u+x|\Lambda|^{-1/2}) - g(a))$$

Since u and a correspond by $u = -g'(a)$ in the duality relation we have as in Lemma 7. $g(a) = s(u) - a \cdot u$ and $a = s'(u)$. Therefore

$$P(X \in dx) \approx \exp|\Lambda|(s(u+x|\Lambda|^{-1/2}) - s(u) - |\Lambda|^{-1/2} x \cdot s'(u)) \approx \exp \frac{x \cdot s''(u) x}{2},$$

so the limiting gaussian should have this density, and hence the covariance matrix $(-s''(u))^{-1}$. To see that this is the same as $g''(a)$ we note that duality correspondence says that the mappings $u = -g'(a)$ and $a = s'(u)$ are inverses of each other. Hence their Jacobians are also inverses of each other, i.e.

$$(\frac{du}{da}) = - g''(a) = (\frac{da}{du})^{-1} = (s''(u))^{-1} .$$

An important feature of this limit theorem is that the limit law is
completely determined by derivatives of $g(a)$ which have a direct
thermodynamic significance and can be experimentally measured. This was
an important insight of Einstein: thermodynamics is also related to
fluctuation phenomena and gives their variances e.g. Brownian motion
or density fluctuations in a gas. To see this relation let us consider
the case $u = (e,n)$ $a = (\beta,-\beta\mu)$, $g = \beta p.$ Then if $g''_{ij}(a) \equiv c_{ij}$
$dg = g'(a)da = - uda$ and $du = - cda$ gives

$$\begin{cases} d(\beta p) = - ed\beta + nd(\beta\mu) \\ de = - c_{11}d\beta + c_{12}d(\beta\mu) \\ dn = - c_{21}d\beta + c_{22}d(\beta\mu) . \end{cases}$$

If we eliminate $d(\beta\mu)$ we get

$$d(\beta\mu) = \frac{e}{n} d\beta + \frac{1}{n} d(\beta p) = \frac{e+p}{n} d\beta + \frac{\beta}{n} dp$$

$$\begin{cases} de = (c_{12} \frac{e+p}{n} - c_{11})d\beta + c_{12} \frac{\beta}{n} dp \\ dn = (c_{22} \frac{e+p}{n} - c_{21})d\beta + c_{22} \frac{\beta}{n} dp \end{cases}$$

Thus we see that e.g.

$$c_{22} \frac{\beta}{n} = (\frac{\partial n}{\partial p})_T = \text{the isothermal compressibility}$$

and c_{22} is the variance of the density fluctuations etc.

Let us also consider the derivation of the famous formula for the motion
of a small but macroscopic brownian particle moving in a fluid or gas. Its
state is determined by (p,q) and its energy is only kinetic $\frac{|p|^2}{2m}$.
It is small compared to that of the whole system, so the probability law
of (p,q) should be given by the exponential law $c \cdot e^{- \frac{\beta|p|^2}{2m}} dpdq$, i.e.
p has a maxwellian distribution, and q has a uniform distribution in
the container. The average motion of the particle can for moderate velocities
and accelerations be described by the equation $\dot{p} = $ the average frictional
force excerted by the medium $= - \alpha v = - \alpha \frac{p}{m}$, where the coefficient is
calculated in hydrodynamics (Stoke's formula) . $\alpha = 6\pi\eta r$ for a sphere

of radius r , where η = the viscosity of the medium. The small random fluctuations in p should then be described by the same equation but with white noise added, $\dot{p} = -\frac{\alpha}{m} p + \sigma\dot{w}$ according to the general philosophy described for the Ehrenfest model in ch. 1. This means that the fluctuations in p should be described by a Gauss-Markov process

$$p(t) = \int_{-\infty}^{t} e^{-\frac{\alpha}{m}(t-s)} \dot{w}(s)ds ,$$ whose equilibrium distribution is gaussian

with variances $\left\langle p_i^2 \right\rangle = \int_0^{\infty} e^{-\frac{2\alpha}{m}t} \sigma^2 dt = \frac{m\sigma^2}{2\alpha}$. This variance has to

agree with the maxwellian one $\frac{m}{\beta}$, which means that σ^2 has to be equal

to $\sigma^2 = \frac{2\alpha}{\beta}$. This is called the fluctuation dissipation relation, because it says that the coefficients σ and α describing these two types of forces have to be related in a specific way. The correlation function of $p_i(t)$ is then given by

$$\left\langle p_i(0) \, p_j(t) \right\rangle = \frac{m}{\beta} \delta_{ij} \, e^{-\frac{\alpha}{m}|t|} ,$$

and since $q(t) - q(0) = \int_0^t \frac{p(s)}{m} ds$ its variance is given by

$$\left\langle (q_i(t) - q_i(0))^2 \right\rangle = \int_0^t \int_0^t \frac{e^{-\frac{\alpha}{m}|u-v|}}{m\beta} du \, dv =$$

$$= \frac{2}{m\beta} \int_0^t du \int_0^u dv \, e^{-\frac{\alpha}{m}(u-v)} = \frac{2}{m\beta} \int_0^t du \, \frac{m}{\alpha} (1 - e^{-\frac{\alpha}{m}u}) \approx \frac{2t}{\alpha\beta} \quad \text{as} \quad \frac{\alpha t}{m}$$

is large. It is in fact not hard to show that as m i small $q(t)$ is approximatively a Wienerprocess governed by the diffusion equation $\frac{\partial P}{\partial t} = \frac{1}{\alpha\beta} \Delta P$.

This is Einsteins celebrated result that the position of the brownian particle is described by a Wiener process with diffusion constant $D = \frac{1}{\alpha\beta} = \frac{kT}{6\pi\eta r}$.

Another example of the fluctuation dissipation relation is the Nyqvist formula for the fluctuating current in an electric circuit. Consider a simple one, e.g. an R-C circuit:

in thermal contact with a heat bath with inverse temperature β. The
state of the system is e.g. described by Q, the charge in the capacitance.
Its energy is $\frac{Q^2}{2c}$, so the exponential probability law is

const $e^{-\frac{\beta Q^2}{2c}}$ giving Q the variance $\frac{c}{\beta}$. The average equations of
motion are given by Ohm's law $\dot{Q} = I = -\frac{V_c}{R} = -\frac{Q}{Rc}$, so the fluctuations
should be described by the Gauss-Markov process defined by $\dot{Q} = -\frac{Q}{Rc} + \sigma\dot{w}$.

This time the F-D relation says that $\frac{\sigma^2 Rc}{2} = \frac{c}{\beta}$, i.e. $\sigma^2 = \frac{2}{\beta R}$. The
equations of motion can be written

$$RI = R\dot{Q} = -\frac{Q}{c} + R\sigma\dot{w} = -V_c + R\sigma\dot{w},$$

which means that the noise can be thought of as generated by a voltage
source $R\sigma\dot{w}$ in series with R:

It has correlation function $(2kTR)\delta(t-s)$, and hence power spectrum
with the constant density $(2kTR)\frac{d\omega}{2\pi}$. This is a general rule for
circuits: each resistance gives rise to a white noise source in series
with it having power density $\frac{kTR}{\pi}$.

References: The thermodynamic limit of entropy etc. is discussed in ref. 9
and 14. The barometric formula is derived in:
C. Marchioro, E. Presutti. Thermodynamics of Particle Systems in the Presence of External Macroscopic Fields. Commun. math. Phys. 27, 146-154 (1972).

The Central Limit Theorem for thermodynamic variables is discussed in:
R.L. Dobrushin, B. Tirozzi. The Central Limit Theorem and the Problem of
Equivalence of Ensembles. Commun. math. Phys. 54, 173-192 (1977),
and in several references given there.

References

1. H.B. Callen. Thermodynamics. Wiley 1960.

2. C. Domb, M.S. Green. Phase Transitions and Critical Phenomena Vol. 1.
    Academic Press 1972.

3. P. & T. Ehrenfest. Begriffliche Grundlagen der statistischen
    Auffassung in der Mechanik. Enc. der Math. Wissenschaften bd. 4,
    Teil 32 (1911).

4. W. Gibbs. Elementary Principles of Statistical Mechanics. Dover 1960.

5. K. Huang. Statistical Mechanics. Wiley 1963.

6. M. Kac. Probability and Related Topics in Physical Sciences.
    Interscience Publ. 1959.

7. A.I. Khinchin. Mathematical Foundations of Statistical Mechanics.
    Dover 1949.

8. L. Landau, E.M. Lifshitz. Statistical Physics. Pergamon Press 1969.

9. O. Lanford. Entropy and Equilibrium States in Classical Statistical
    Mechanics. In Statistical Mechanics and Mathematical Problems.
    Springer Lecture Notes in Physics 20. Springer 1973.

10. R.A. Minlos. Lectures on Statistical Physics. Russian Mathematical
    Surveys 23 (1968) no. 1.

11. A.B. Pippard. Classical Thermodynamics. Cambridge University Press 1957.

12. C. Preston. Random Fields. Springer Lecture Notes in Mathematics 534.
    Springer 1976.

13. R.T. Rockafellar. Convex Analysis. Princeton University Press 1970.

14. D. Ruelle. Statistical Mechanics. Benjamin 1969.

15. G.E. Uhlenbeck, G.W. Ford. Lectures in Statistical Mechanics.
    American Mathematical Society 1963.

Texts and Monographs in Physics

Editors: W. Beiglböck, M. Goldhaber,
E. H. Lieb, W. Thirring

A. Böhm
Quantum Mechanics
1979. 105 figures. Approx. 570 pages.
ISBN 3-540-08862-8

O. Bratelli, D. W. Robinson
Operator Algebras and Quantum Statistical Mechanics
Volume 1
The Mathematical Theory of C*- and W*-Algebras
1979. Approx. 500 pages.
ISBN 3-540-09187-4

H. Pilkuhn
Relativistic Particle Physics
1979. Approx. 400 pages.
ISBN 3-540-09348-6

R. D. Richtmyer
Principles of Advanced Mathematical Physics I
1978. 45 figures. XV, 422 pages.
ISBN 3-540-08873-3

R. M. Santilli
Foundations of Theoretical Mechanics I:
The Inverse Problem in Newtonian Mechanics
1978. 5 figures. IX, 266 pages.
ISBN 3-540-08874-1

M. D. Scadron
Advanced Quantum Theory of Its Applications through Feynman Diagrams
1979. 78 figures. Approx. 300 pages.
ISBN 3-540-09045-2

J. Kessler
Polarized Electrons
1976. 104 figures. IX, 223 pages.
ISBN 3-540-07678-6

W. Rindler
Essential Relativity
Special, General, and Cosmological
Second Edition 1977. 44 figures.
XV, 284 pages.
ISBN 3-540-07970-X

K. Chadan, P. C. Sabatier
Inverse Problems in Quantum Scattering Theory
1977. 24 figures. XXII, 344 pages.
ISBN 3-540-08092-9

J. M. Jauch, F. Rohrlich
The Theory of Photons and Electrons
The Relativistic Quantum Field Theory of Charged Particles with Spin One-Half
Second Expanded Edition 1976.
55 figures, 10 tables. XIX, 553 pages
ISBN 3-540-07295-0

C. Truesdell, S. Bharatha
The Concepts and Logic of Classical Thermodynamics as a Theory of Heat Engines
Rigorously Constructed upon the Foundation
Laid by S. Carnot and F. Reech
1977. 15 figures. XXII, 154 pages.
ISBN 3-540-07971-8

Selected Issues from

Lecture Notes in Mathematics

Lecture Notes in Physics

This series reports new developments in physical research and teaching – quickly, informally and at a high level. The type of material considered for publication includes:

1. Preliminary drafts of original papers and monographs
2. Lectures on a new field or presentations of new angles in a classical field
3. Seminar work-outs
4. Reports of meetings, provided they are
    a) of exceptional interest and
    b) devoted to a single topic.

Texts which are out of print but still in demand may also be considered if they fall within these categories.

The timeliness of a manuscript is more important than its form, which may be unfinished or tentative. Thus, in some instances, proofs may be merely outlined and results presented which have been or will later be published elsewhere. If possible, a subject index should be included. Publication of Lecture Notes is intended as a service to the international physical community, in that a commercial publisher, Springer-Verlag, can offer a wide distribution of documents which would otherwise have a restricted readership. Once published and copyrighted, they may be documented in the scientific literature.

Manuscripts

Manuscripts should be no less than 100 and preferably no more than 500 pages in length.

They are reproduced by a photographic process and therefore must be typed with extreme care. Symbols not on the typewriter should be inserted by hand in indelible black ink. Corrections to the typescript should be made by pasting in the new text or painting out errors with white correction fluid. Authors receive 50 free copies and are free to use the material in other publications. The typescript is reduced slightly in size during reproduction; best results will not be obtained unless the text on any one page is kept within the overall limit of 18 x 26.5 cm (7 x 10½ inches). On request, the publisher will supply special paper with the typing area outlined.

Manuscripts in English, German or French should be sent to Prof. Dr. W. Beiglböck, Institut für Angewandte Mathematik, Im Neuenheimer Feld 5, 6900 Heidelberg/Germany, or directly to
Springer-Verlag Berlin Heidelberg GmbH

Springer-Verlag Berlin Heidelberg GmbH

ISBN 978-3-540-09255-1   ISBN 978-3-540-35293-8 (eBook)
DOI 10.1007/978-3-540-35293-8